Partnerships for Empowerment

Dedicated to the memory of Jan Lowrey (1949–2006),
colleague, community partner, visionary, mentor, friend.

Partnerships for Empowerment

Participatory Research for Community-based Natural Resource Management

Edited by Carl Wilmsen, William Elmendorf, Larry Fisher, Jacquelyn Ross, Brinda Sarathy and Gail Wells

London • Sterling, VA

First published by Earthscan in the UK and USA in 2008

Paperback ISBN-13: 978-1-84407-563-8
Hardback ISBN-13: 978-1-84407-562-1
Typeset by JS Typesetting Ltd, Porthcawl, Mid Glamorgan
Printed and bound in the UK by Cromwell Press, Trowbridge
Cover design by Susanne Harris

For a full list of publications please contact:

Earthscan
Dunstan House, 14a St Cross Street
London, NW1 0JH, UK
Tel: +44 (0)20 7841 1930
Fax: +44 (0)20 7242 1474
Email: earthinfo@earthscan.co.uk
Web: **www.earthscan.co.uk**

22883 Quicksilver Drive, Sterling, VA 20166-2012, USA

Earthscan publishes in association with the International Institute for Environment and Development

A catalogue record for this book is available from the British Library

Library of Congress Cataloging-in-Publication Data

Partnerships for empowerment : participatory research for community-based natural resources management / edited by Carl Wilmsen ... [et al.].
p. cm.
ISBN 978-1-84407-563-8 (pbk.) – ISBN 978-1-84407-562-1 (hbk.)
1. Natural resources–Co-management. 2. Sustainable forestry. I. Wilmsen, Carl, 1956–
HC85.P37 2008
333.7–dc22

2008011462

The paper used for this book is FSC-certified. FSC (the Forest Stewardship Council) is an international network to promote responsible management of the world's forests.

Mixed Sources
Product group from well-managed forests and other controlled sources
www.fsc.org Cert no. TT-COC-2082
© 1996 Forest Stewardship Council

Contents

List of Figures, Tables and Boxes

Figures

TABLES

BOXES

List of Contributors

Heidi L. Ballard is an assistant professor in the School of Education at the University of California at Davis. Her research and teaching focuses on conservation issues, participatory research and participatory biodiversity monitoring.

Kathleen Bond is a professional facilitator/mediator focusing on natural resource/environmental conflict resolution. Casework includes facilitating a multi-year, community-based, landscape-scale collaborative process for forest plan revision for the Grand Mesa, Uncompahgre and Gunnison National Forests in western Colorado.

Shannon Brawley is project coordinator of the Tending and Gathering Garden at the Cache Creek Nature Preserve in Woodland, California. She received her undergraduate degree in landscape architecture and has continued on as a graduate student in the Geography Graduate Group at the University of California, Davis. Her research interests are in the areas of ecosystem restoration, participatory research and community-based resource management.

Mae Burnette cares for springs in her duties as restoration projects supervisor for the White Mountain Apache Tribe. She was formerly a crew boss for fire fighters. She lives in Whiteriver, Arizona.

Antony (Tony) S. Cheng is an assistant professor of forestry and natural resource policy in the Department of Forest, Rangeland and Watershed Stewardship at Colorado State University, Fort Collins, Colorado.

Don Collins is the president of the Northwest Research and Harvesters Association, which operates throughout Washington and Oregon. He has been harvesting non-timber forest products for over 50 years, and is a retired employee of the US Department of Agriculture (USDA) Forest Service out of Olympic National Forest in Washington State.

Gabriel Cumming is a PhD candidate in ecology at the University of North Carolina, Chapel Hill. His research employs documentary ethnography to examine ecological discourses and community-based natural resource management strategies in rural North Carolina. Gabriel is from Greenwood, South Carolina, and received his BA from Swarthmore College.

Brian W. Eisenhauer serves as the associate director of the Center for the Environment and as assistant professor of sociology in the Social Sciences Department at Plymouth State University. His primary area of specialty is natural resource/ environmental sociology, a field that studies community in its largest sense by examining the interrelationships between society, culture and the environment.

William F. Elmendorf is an assistant professor and the extension specialist for urban and community forestry in the School of Forest Resources at the Pennsylvania State University. He teaches community forest management at Penn State and provides technical assistance and education to Pennsylvania municipalities, agencies and volunteer organizations. His research interests are social in nature and he continues to complete survey work on ethnic groups, land-use policy and recreation.

Delbin Endfield is a project manager with the White Mountain Apache Tribe who has coordinated community-based watershed restoration activities in Cibecue, Arizona, since 1997. He resides in Canyon Day, Arizona.

Larry Fisher is senior program manager at the US Institute for Environmental Conflict Resolution in Tucson, Arizona. He holds a PhD in forest policy and management from Cornell University and has extensive experience with participatory research in Indonesia.

Stephanie L. Gripne, PhD, is a land conservation program manager for The Nature Conservancy of Colorado. Her work focuses on conservation finance, conservation real estate and economic strategies that provide non-market goods and services.

Stacy Guffey is the ninth generation of his family to call Macon County, North Carolina, home. As the county's planning director, he emphasizes the importance of meaningful public participation when making decisions about the future. A community leader on issues of planning and conservation, he earned an undergraduate degree from Western Carolina University.

Don L. Hankins is an assistant professor in geography and planning at California State University, Chico. He has conducted participatory research among his own and other indigenous communities within California and abroad. He views participatory research as a tool to achieve resource management and conservation.

Susan Hansen has been Delta County, Colorado, administrator since 1992. She also serves as facilitator for the North Fork Coal Working Group, a collaborative group of government, business and environmental group representatives that meets quarterly to share information about, and address potential impacts of, coal development activities in Delta and Gunnison counties. She and her husband own and operate a cattle ranch in Crawford, Colorado.

Ajit Krishnaswamy is socio-economic extension specialist in the School of Resource and Environmental Management, Simon Fraser University, Canada. He is the former director of the National Community Forestry Center, a project of the US-based National Network of Forest Practitioners (NNFP). He worked for the Indian Forest Service for several years before coming to North America. He has held positions as the president of the Institute for Culture and Ecology located in Portland, Oregon, and the project manager of the World Commission on Forests and Sustainable Development at the International Institute of Sustainable Development, Winnipeg, Canada.

Carmine Lockwood is the planner for a large national forest headquartered in Delta, Colorado. His background includes 25 years of experience serving in numerous Forest Service positions throughout the western US. He is now leading the forest plan revision effort. Working on seven forests in three regions, he has become a practitioner of collaborative approaches to working through conflicts during project and program planning.

Jonathan Long is an ecologist who has worked in mountain and canyon regions of the southwestern US. He has worked for the White Mountain Apache Tribe and has served as an extension agent for the Hualapai Tribe in Arizona. He currently coordinates research projects for the Pacific Southwest Research Station in Lake Tahoe. His professional interests include the eco-cultural restoration of wetlands and watersheds, and strategies for communicating ecological change.

Jan Lowrey was the executive director of the Cache Creek Conservancy in Yolo County, California. He was a fourth generation farmer/land manager with a background in riparian restoration, native species re-vegetation and gaining land-owner cooperation. Jan passed away unexpectedly in January 2006.

Candy Lupe was formerly the director of the White Mountain Apache Tribe's Watershed Program and coordinator of the White Mountain Apache Conservation District. She resides in Whiteriver, Arizona.

Carla Norwood is a graduate student in ecology at the University of North Carolina at Chapel Hill. She grew up on an old tobacco farm in Warren County, North

Carolina, and earned both an undergraduate degree and an MSc in environmental management from Duke University. She works with communities to analyze changing development patterns and harness geospatial tools to serve the public good.

Michael Rios is assistant professor in the Department of Environmental Design at the University of California, Davis. Previously he held a joint faculty appointment in the School of Architecture and Landscape Architecture at Pennsylvania State University, where he was also inaugural director of the Hamer Center for Community Design from 1999 to 2007. His research interests focus on the analysis and assessment of contemporary public policy, professional practice and citizen participation in the planning and design of public landscapes.

Jacquelyn Ross is assistant director for immediate outreach for the Undergraduate Admissions and Outreach Services at the University of California, Davis. She also acts as a consultant to the California Indian Basketweavers Association and the Tending and Gathering Garden at the Cache Creek Conservancy.

Brinda Sarathy is an assistant professor in environmental studies and international/intercultural studies at Pitzer College in Claremont, California.

Joyce A. Trettevick is a Makah tribal member and has been helping to manage the forests of the 30,000 acre (12,140 hectare) Makah Indian Reservation for over 30 years.

Johanna M. Ward has a PhD in ecology and is the regional scientist for the Rocky Mountain Conservation Region of The Nature Conservancy, where she works on cross-boundary priorities focused on grassland and arid land conservation. She spends most of her time working with federal land agencies to improve the management of public lands that dominate the interior west.

Gail Wells is a writer, editor and communications consultant. She has published extensively on sustainability, as well as on the development of the state of Oregon's forests. She has had numerous editorial positions, including director of the Forestry Communications Group at Oregon State University.

Carl Wilmsen, PhD, is a specialist in community forestry at the University of California, Berkeley, where he directs the Community Forestry and Environmental Research Partnerships (CFERP). He has published journal articles and book chapters on institutionalized racism and natural resources, forest land-use issues, participatory research and oral history.

J. D. Wulfhorst is director at the Social Science Research Unit and associate professor of rural sociology at the University of Idaho. His areas of research and scholarship focus on the social impacts of agricultural production and natural resource uses in rural community settings.

Acknowledgments

Many minds and hands touched this book as we moved it from an idea to words on the printed page. The book originated as an idea of the steering committee of the Community Forestry and Environmental Research Partnerships (CFERP) (formerly the Community Forestry Research Fellowships Program). CFERP and other organizations, such as the National Network of Forest Practitioners (NNFP), had been using participatory research to advance community forestry and community-based natural resource management in the US for many years, but no critical assessment of the accumulating body of experience had occurred. As a way to begin assessing the impacts and continuing challenges in this area of endeavor, CFERP convened several graduate student fellows, community partners, scholars and professionals who use participatory research at a workshop in 2003. The workshop and subsequent discussions deepened our appreciation of the kinds of impacts participatory research can have, as well as of the challenges, frustrations and hard work that it takes in using it in pursuit of a more sustainable, equitable and joyful world.

The CFERP annual workshops also increased our appreciation of the strengths and limitations of participatory research and contributed a great deal to the development of the ideas in the chapters in this volume. We are grateful to the many community partners, faculty advisers, graduate student fellows and others who shared their insights, experience and knowledge over the years. We are also grateful to the Ford Foundation for its strong long-term support of CFERP and its financial underwriting of the 2003 workshop.

Although the people who contributed insights, thoughts and ideas are too numerous to thank individually, a few people deserve special mention. First, John Bliss, Louise Fortmann, Tamara Walkingstick, and Dreamal Worthen all reviewed drafts of chapters and helped the authors to develop their ideas. They, together with Larry Fisher, were members of the CFERP steering committee in 2003 and helped to make the workshop a lively, stimulating experience. We are also very grateful to Kelly Perce for her expert management of the numerous drafts of the chapters, her skill in ensuring that they were all formatted correctly and her careful attention to the flow of logic and readability in the draft chapters that she reviewed.

Carl Wilmsen, William Elmendorf, Larry Fisher, Jacquelyn Ross,
Brinda Sarathy and Gail Wells
February 2008

Foreword

Academic research seeks to answer questions, to test hypotheses and to reach for deeper understandings of people, places and processes. Applied research seeks to find solutions for real world problems and research on rural community development usually tries to be truly relevant to specific situations. However, most research invariably begins with someone asking *why*? Too often, the question of *who* is asking, who has framed the problem and who really needs to know the answer is assumed. Usually it is the researchers themselves, sometimes a policy-maker or a manager who frame the question – but rarely a poor forest-dependent logger, a tribal elder, a community-based entrepreneur or an informal collaborative of community stakeholders trying to revive the rural economy that forms their identity and is the source of their livelihoods. When research is derived and driven by locally articulated questions and needs, the act of formulating methodology, collecting data and, especially, analyzing the information gathered to draw conclusions is still invariably done by an external agent, someone outside the concerned community. When this happens, ownership of the process, the information and the results is rarely internalized by stakeholders most directly affected and able to use the knowledge produced. Research is most often done *to*, *for* or *about*, or even *on behalf of* rural communities and is occasionally *bestowed upon* them for their use. Even the best-intentioned academic efforts can emit a pungent odor of arrogance to the rest of society. After all, research – like most other activities that we humans engage in – is about pride and power. And, like most efforts to share power and act with humility, sharing research in all its aspects is rarely easy. It rattles assumptions about who is the researcher and who can do research. In addition, it is almost always messy. And academic rigor does not particularly relish messiness.

Fortunately, in the field of rural development and community-based natural resource management many good thinkers have been trying to develop methodologies, modify mindsets and reduce the messiness of sharing the design and application, the conduct and the use of research. Much of this really important work on collaborative and participatory research has been pioneered outside of the US: indeed, 'participatory' approaches have become so dominant in international development circles as to be seen as a new orthodoxy. I have been in villages in India that have become field training sites for participatory rural appraisal (PRA), where villagers patiently instruct researchers on the need to do a participatory community profile before a wealth ranking, and a Venn diagram of intersecting institutions before a gendered matrix ranking of tree species: 'As you ask us to participate, please get it right!' Working honestly together to try and 'get it right' in the US context is what this much needed book is all about.

Building on ten years of hard work by scores of Community Forestry Research fellows and collaborating communities across the country, this book looks at *how* we can try to do research in a humble, sharing and effective manner without glossing over how difficult it is to do in practice. Both 'academic' and 'community' researchers need support and tools in this process and there are many pitfalls along the way. The book explores both the agonies and the ecstasies of trying to engage in truly collaborative research. It discusses hard work that lies at the interface of different learning styles and worldviews. It addresses the problem of blending scientifically procured, deductive knowledge and the received wisdom of traditional knowledge, and looks at trade-offs between rigor and intuition. The book explores the contradictory role of the researcher when she or he is engaged in action, becoming a part of the community as opposed to the more typical 'objective and dispassionate' recorder of information. One real difficulty in collaborative research is the asymmetry of existing understanding and awareness. Often, the academic researcher benefits the most – gaining rare insights into a community and its complex context. For the community it is frequently an altruistic effort to help the researcher understand what everyone else already knows – to 'bring them up to speed' – and this can be time consuming, frustrating and one sided in terms of benefits. Furthermore, the academic researcher is often most useful to the community when the research is completed and follow-up is needed to apply the solutions. This is, sadly, when the researcher usually leaves to complete a degree or take up a job elsewhere, repeatedly leaving a community frustrated and feeling abandoned. At its best, participatory research can be truly empowering – an interpretative or transboundary communication vehicle allowing the community to use the tools of the powerful and speak with an externally acceptable voice. Co-option can go both ways.

Many of the ideas in this book were first explored by academic and community research partners and their academic advisers in the annual workshops of the Community Forestry Research Fellowship program, some of which I had the pleasure of attending. It is a testament to the integrity, honesty and continuous efforts of all of these people that they persevered in their pursuit of collaborative and participatory methods in spite of many obstacles. Ironically, it may be academia itself, particularly in the US, which presents the greatest resistance to the innovative approaches discussed in this excellent book. As the fellows of the CFRF program infiltrate the ivory towers with some of these messy heretical methodologies, and experienced communities see the value of partnering proactively with universities, community colleges and research institutes, I hope that a new paradigm of truly relevant and empowering research does, indeed, become a new 'orthodoxy' in the US as well.

Jeffrey Campbell
Senior Program Officer
Community and Resource Development, The Ford Foundation
February 2008

List of Acronyms and Abbreviations

AISES	American Indian Science and Engineering Society
AR	action research
BAC	Bioregional Advisory Council
BelCUP	Belmont Community University Partnership
CBNRM	community-based natural resource management
CCDRE	Committee for Community Directed Research and Education
CCNP	Cache Creek Nature Preserve
CCRMP	Cache Creek Resources Management Plan
CFERP	Community Forestry and Environmental Research Partnerships
CFRF	Community Forestry Research Fellowships
CIBA	California Indian Basketweavers Association
the Committee	TGG Steering Committee
the Conservancy	Cache Creek Conservancy
CSREES	Cooperative State Research Extension and Education Service
DNR	US Department of Natural Resources
FRP	Friends Rehabilitation Program
the Garden	Native American Tending and Gathering Garden
GIS	geographic information systems
GMUG	Grand Mesa, Uncompahgre and Gunnison National Forests
GREM	grassroots ecosystem management
ICE	US Immigration and Customs Enforcement
IRB	Institutional Review Board
km	kilometer
LFWLG	Latino Forest Workers Leadership Group
LTP	Little Tennessee Perspectives
LWG	Landscape Working Group
m	meter
MCBFI	Makah Community-based Forestry Initiative
n	total sample population size
NAC	National Advisory Council

NCFC	National Community Forestry Center
NGO	non-governmental organization
NNFP	National Network of Forest Practitioners
NRHA	Northwest Research and Harvester Association
NTFP	non-timber forest product
NTFR	non-timber forest resource
PAR	participatory action research
PLA	participatory learning and action
PR	participatory research
PRA	participatory rural appraisal
PWCFC	Pacific West Community Forestry Center
RRA	rapid rural appraisal
SRO	single-room occupancy (hotels)
TEK	traditional ecological knowledge
TGG	Tending and Gathering Garden
UC Davis	University of California at Davis
UNC	University of North Carolina
UNDP	United Nations Development Programme
US	United States
USDA	United States Department of Agriculture
USFS	United States Forest Service
VISTA	Volunteers in Service to America
WA DNR	Washington State Department of Natural Resources

1

Negotiating Community, Participation, Knowledge and Power in Participatory Research

Carl Wilmsen

Here come the anthros, better hide your past away. (Floyd Red Crow Westerman)

Introduction

In 1970 actor and singer Floyd Red Crow Westerman released a record album named after Vine Deloria Jr's controversial book *Custer Died for Your Sins*. One song on the album, 'Here Come the Anthros', wryly takes anthropologists to task for misrepresenting and disrespecting Native American peoples while simultaneously failing to help them retain their cultures and revitalize their communities. Deloria and Westerman were not alone in finding fault with traditional approaches to research. Theirs were among the many critiques that began to emerge during the 1960s and 1970s of standard research practice, economic development and conservation. Academic researchers, development practitioners and conservationists working in Africa, Asia and Latin America during that time period began to argue that the 'top-down' prescriptions for economic development of outside experts were failing to alleviate poverty, create greater social equality and halt environmental degradation.

To fix this problem, they called for greater community participation in conservation and economic development. They reasoned that this approach would yield more successful outcomes because new development institutions, decision-making processes and community assets would be grounded in local needs and realities. To ensure such grounding, insider knowledge of local customs, mores and political-economic context, as well as detailed local environmental knowledge were needed. The stage was thus set for adopting participatory approaches to research in natural resource issues.

A decade or two later this same reasoning was applied to the US. For too long, the prescriptions for community development and the conservation of natural resources of outside experts, whether academics, professional consultants, industrialists or government bureaucrats, have produced outcomes with questionable benefits for local communities. Calls for participatory approaches to research have accompanied calls for collaborative approaches to community development and natural resource management. Participatory research has thus emerged as an approach to producing knowledge that is sufficiently grounded in local needs and realities to support community-based natural resource management in the US, and it is often touted as crucial to the sustainable management of forests and other natural resources (Gray et al, 2001; Baker and Kusel, 2003).

Are these claims justified? Recent scholarship suggests that participatory research (PR) does not always meet the mark. Scholars have pointed out the difficulty that communities encounter in reaching out to disenfranchised community members (Schafft and Greenwood, 2002), have suggested that PR may be used in ways that exclude community members from decisions about how research results are applied (Simpson, 2000), and have argued that often what is called participatory research is nothing more than 'contracting people into projects which are entirely scientist led, designed and managed' (Cornwall and Jewkes, 1995, p1669).

Community-based natural resource management (CBNRM)[1] suffers from these same problems. For example, in a theme review paper that he wrote for a workshop on community-based conservation in 1993, anthropologist Marshall Murphree observed that out of 15 case studies presented at the workshop, only two were actually conceived and initiated by communities. Murphree (1993)suggested that this may have occurred because governments, interest group organizations and scholars are typically the ones who define what constitutes conservation. Thus, efforts to involve communities in conservation are efforts to co-opt community support for objectives that originate elsewhere (Murphree, 1993). Similarly, Cooke and Kothari's (2001) review of recent experience with participation in development leads them to wonder whether participation has actually become a new orthodoxy that does more to maintain inequities in access to resources and political power than it does to empower community members.

What does this mean for adopting participatory approaches to research in community-based natural resource management in the US? Can such approaches lead to more meaningful participation, community capacity-building and the sustainable management of natural resources?

This book addresses these questions by critically analyzing case studies of distinct recent experiences and by discussing critical issues. In compiling and comparing these cases and essays, all but two of which were written by teams of scholars and practitioners directly involved, the book identifies the unique features of interweaving PR and CBNRM under contemporary economic, political and environmental conditions in the US. It outlines the continuing challenges and ongoing issues and draws lessons from them that are applicable to CBNRM the world over. The book's objective is to assess whether and how participatory approaches to research can help to achieve the CBNRM goals of developing communities through empowering them to manage resources sustainably.

The experience contained in these case studies suggests that PR can contribute to these goals, but that doing so involves much careful negotiation among research collaborators over a number of issues. The practice of PR offers no guarantee that the goals of more meaningful participation, capacity-building and the democratization of natural resource management will be met. There are many ongoing issues in participatory research, including the problematic nature of participation, balancing rigor and relevance, addressing power relations, and, indeed, the very notion of community itself. All or some of these issues will arise under unique circumstances in every PR project. Dealing with them requires measures that the research collaborators negotiate and tailor to the specifics of each situation. While PR practitioners have developed a means of addressing many of these issues, there are no hard and fast rules, answers or techniques that apply in every case. One size definitely does not fit all in participatory research.

The authors of the chapters of this book discuss these issues as they encountered and dealt with them. Their approaches illustrate how practitioners in CBNRM are working to make research more relevant to communities while simultaneously producing robust understandings of the world and how it functions. Their work, with its successes and mistakes, suggests that community empowerment through participatory research is a work in progress. Every PR project is embedded within a specific set of political-economic relations, both internal and external to the community, that hinder or facilitate achieving empowerment goals, often doing both simultaneously. Yet, it was the bearing of political-economic relations on traditional scientific research and conservation projects that led to participatory research in the first place. Why? What was the reasoning that brought us to this point in history when the ideas and issues in this book are being thought about and discussed? Tracing the history of the development of PR in CBNRM shows how the studies in this volume are themselves historically situated. That is to say, it reveals the issues that have led scholars to adopt participatory approaches to research, the many streams of thought that have shaped participatory approaches, and the implications that these issues and streams of thought have for our very understanding of how we can learn about the world. This new way of understanding how we can learn about the world, the new epistemology, is crucial to creating conditions in which ordinary people are positioned to benefit directly from research and use research results to improve their own lives or have a voice in their own affairs. While it may be a necessary condition to empowering communities, it is by no means sufficient, as the chapters in this volume demonstrate.

THE RISE OF PARTICIPATORY RESEARCH IN NATURAL RESOURCE MANAGEMENT IN THE US

By the 1960s it was becoming increasingly clear that something was wrong. The centrally planned capital-intensive development aid programs championed by international development institutions such as the World Bank and the United Nations Development Programme (UNDP), as well as by industrialized nations, were failing to alleviate poverty and reduce income disparity in developing countries. There was also a parallel critique of traditional, positivist approaches to scientific research that was in full swing by the 1970s. In this context, professionals in planning, industry, education, public health, natural resources and other fields began to search for more democratic forms of management. They sought an end to 'top-down', expert-driven technocratic approaches to the many problems that communities around the world faced. At the same time they sought ways of learning about the conditions under which ordinary people actually lived that would directly support new, more equitable forms of management. This search led to the development of community-based approaches to environmental management and economic development. It also led to the adoption of participatory approaches to research in natural resources. One particularly influential approach to learning about the everyday realities of rural people and the conditions under which they live and make a living that emerged during the late 1970s was rapid rural appraisal (RRA). Like community-based conservation, and in conjunction with grassroots development, RRA grew out of dissatisfaction with the status quo in research practice as applied to standard rural community development (see Box 1.1 for definitions). Such practice was criticized for three major reasons. First, it tended to focus investigations on officially sanctioned projects in easily accessible areas (to the neglect of peripheral areas), on the experiences of men (thereby excluding or minimizing the experiences of women) and on the experiences of the elite (excluding the poor). Second, it often relied on questionnaire surveys that were found to produce inaccurate and unreliable data, and to result in reports that were not useful and usually ignored. Third, it was expensive (Chambers, 1994).

In response to this criticism, RRA was developed as a set of approaches and methods for quickly and cost effectively learning about rural livelihood practices, the factors that impede or support them, and their environmental and social implications.[2] RRA practitioners drew on a variety of sources, including approaches that emphasized participation to varying degrees and that were grounded in different ideologies. These included approaches that viewed farming as a complex system of human and non-human components (farming systems research), approaches that analyzed the ecology of agricultural systems (agro-ecology), and approaches that applied the insights of anthropology to solving practical problems (applied anthropology).

Box 1.1 Definitions of community-based approaches to environmental management and economic development

Community-based conservation is the protection of biodiversity and natural resources in collaboration with community groups. Other collaborators may include government agencies and environmental groups. Rather than outside experts working to protect nature without any community input, help or consent (as in standard 'top-down' approaches), community members share in decisions about, and management of, protected areas. Communities are typically interested in regaining control of the land being protected and in improving their economic situation through conservation-related activities such as guide services (Western and Wright, 1994).

Community-based natural resource management (CBNRM) is careful management of timber harvesting, non-timber forest products (mushrooms, berries, materials for crafts, and others) gathering, cattle grazing, fishing and other extractive activities to ensure the ecological sustainability of the resources while simultaneously improving community well-being. It is typically a collaborative process involving community leaders, regional and/or national government officials, non-governmental organizations (NGOs) and others in shared decision-making and policy formation. It may entail devolving authority over natural resources to local governments. Different countries around the world have devolved authority to different degrees.

Rural community development is purposeful effort to improve a community's well-being. Traditionally, community development has been approached more in terms of *economic development* in which the creation of new businesses and jobs is emphasized. A broader approach that is often adopted in community-based efforts entails more than just improvement in economic conditions. It also entails equitable access to resources, as well as distribution of the costs and benefits of development. In addition, it entails building the capacity of community members to work together in addressing their environmental, social and political interests and concerns (Wilkinson, 1991).

Some of these approaches drew on the work of Karl Marx – who argued that capitalism is oppressive to workers because workers do not own the means of production (money and raw materials) or the products of their labor – for inspiration and theoretical understanding. Others were grounded in the work of pragmatist philosophers such as John Dewey, who held that action and knowledge are inseparable, and who sought a stronger democracy through the participation of all levels of society (Greenwood and Levin, 1998). Robert Chambers' writing has been especially influential in popularizing the rural appraisal approach, although he is careful to acknowledge that many people and institutions contributed to the development of rapid rural appraisal (RRA) (Chambers, 1994, p956).

Chambers (1994) has also noted, however, that while RRA affords community members greater engagement in the research process than traditional questionnaire surveys, it is a technique that is designed primarily for outsiders to learn. During the late 1980s, participatory rural appraisal (PRA) began to emerge as a means of empowering local communities to conduct their own analyses. In this approach, the main goal is for communities themselves to learn and gain knowledge that they can then apply to ameliorating their own problems. The professional researcher acts only as a facilitator and catalyst.

In the mid 1990s, interest in PRA blossomed. Now often being called participatory learning and action (PLA), PRA/PLA and other approaches spread to all corners of the globe. Aid agencies began to require participation (often PRA) in their projects (Cooke and Kothari, 2001; Chambers, 2005), and participatory approaches to research in natural resource issues spread from the global South (i.e. developing countries) to the global North (i.e. industrialized countries) (Flower et al, 2000).

To be sure, participatory approaches were already being used in the US. John Gaventa, for example, took a participatory approach to his study of power relations in coal-mining communities in Appalachia (Gaventa, 1980). The Highlander Research and Education Center in Tennessee had been using participatory techniques to address civil rights and social justice issues for decades. In addition, urban and community planners had been using participatory techniques and methods since at least the 1960s (Sanoff, 2000).

However, interest in adopting participatory approaches in the US increased during the mid to late 1980s. At that time, people who had worked abroad in community-based development and conservation began to apply their overseas experience to the problems of resource-dependent communities at home. Some communities also began to explore collaborative approaches to resolving their conflicts over natural resource use and management. These scholars and practitioners realized that although the context was different, the issues such communities faced were very similar, if not the same, as those confronted by their counterparts overseas: communities were effectively barred from participating in natural resource management decisions that directly affected their livelihoods; they needed access to resources, such as capital, information and raw materials, to maintain or improve their livelihood practices; they continued to struggle with high poverty rates and, in many cases, economic restructuring of natural resource industries, while efforts to preserve biological diversity exacerbated the situation. Reflecting on the experience overseas, community-based development practitioners wondered whether conservation that simultaneously maintained healthy ecosystems and healthy communities could be achieved in the US.

Very soon thereafter, some of the first collaborative efforts in natural resource management began to emerge. As these efforts unfolded, communities sought research that supported their community-based sustainable development efforts (Baker and Kusel, 2003). This led to a burst of interest and activity in participatory research. By the late 1990s, PR was gaining ground as an approach to studying natural resource management issues in the US.

DEMOCRATIZATION OF RESEARCH AND SOCIETY THROUGH PARTICIPATORY RESEARCH

Today, participatory approaches to research in US settings have moved beyond RRA and PRA/PLA to incorporate insights, ideas, methods and techniques from many other fields with participatory traditions of their own. Like their counterparts in natural resources, practitioners and scholars in community planning, education, industry and public health have been concerned since at least the 1960s and 1970s that expert-driven technocratic approaches in their fields led to or exacerbated social inequities and prevented the achievement of the full potential of production processes or human capabilities. They borrowed from each other's traditions and often grounded their work in the same philosophical traditions. Depending upon the political orientation of the practitioners involved, participatory practice in any of these traditions could be rooted in Marxist, pragmatist or other political-economic philosophies (see Table 1.1). PR practitioners in natural resources in the US have borrowed freely from these traditions.

While each of these fields has contributed to the search for more equitable forms of natural resource management and better research methods for learning about use and engagement with land and natural resources, Kurt Lewin, in social psychology, and Paulo Freire, in education, generated two traditions that have been especially influential in the development of participatory approaches to research in natural resource management (Freire, 1981). Kurt Lewin originated action research (AR) in his work in the industrial democracy movement of the 1940s and 1950s (Lewin, 1948). Lewin conceptualized technological and social systems as interlinked and interdependent. Following Dewey and other pragmatists, he averred that knowledge is produced through action and that workers have knowledge of production practices through their intimate involvement with them on the factory floor. This reasoning led to the training of workers to contribute to innovations in production processes, as well as a concern for the democratization of production practices. By involving workers in solving problems of production, industry could tap into workers' knowledge and use it as a foundation for sound decisions (Greenwood and Levin, 1998).

Action research thus had two interconnected goals. On the one hand, the goal was to produce a better account of the world that could be used for practical problem-solving through the democratization of production processes. On the other hand, Lewin's ideas also applied to a social change project: democratizing society through the democratization of the workforce. While this more radical goal faded from the industrial democracy movement – indeed, action research can be co-opted by industry for economic rather than democratic objectives – AR maintains its democratizing goals and remains grounded in the thought of pragmatist philosophers and reformers such as John Dewey, Richard Rorty, Charles Pierce and William James (Greenwood and Levin, 1998).

Table 1.1 *Participatory approaches to research*

Methodological approach	*Field of application*	*Date of origin*	*Origins*	*Background influences*[a]
Action research	Industrial democracy; many others today	1940s	Kurt Lewin (1948)	John Dewey; pragmatism
Participatory planning	Planning	1960s	Robert Goodman (1972); John Friedman (1973); John Forester (1989)	John Dewey; pragmatism; Karl Marx; Kurt Lewin; Paulo Freire
Conscientization[b]	Popular education	1970	Paulo Freire (1981)	Karl Marx
Rapid rural appraisal (RRA); participatory rural appraisal (PRA)/ participatory learning and action (PLA)	Grassroots development and community-based conservation	1970s and 1980s	Disparate origins in farming systems research; applied anthropology; agro-ecosystem analysis; development studies Robert Chambers (1980)	Paulo Freire; Kurt Lewin; Karl Marx
Participatory action research (PAR)	Community-based natural resource management (CBNRM); many other fields	1980s and 1990s	Disparate origins in action research and conscientization	John Dewey; pragmatism; Karl Marx; Kurt Lewin; Paulo Freire; post-Marxist studies; feminism
Community-based participatory research	Health	1990s	Many	Conferences on participation in health in 1977, 1986 and 1996; Paulo Freire; post-Marxist studies; feminism

Note: [a] Many more people and schools of thought have been influential in each of these fields than can neatly fit into this table. Just some of the major figures are listed here to show the disparate and overlapping origins of participatory traditions in these fields.
[b] Conscientization: students and teachers learning together about the roots of oppression by synthesizing the knowledge each brings to the discussion.

Paulo Freire's work is similarly aimed at creating a more robust and inclusive democracy. His influential book *Pedagogy of the Oppressed* (Freire, 1981) makes many of the same claims about knowledge production as the pragmatist philosophers and Kurt Lewin: knowledge is gained through action, workers have intimate, detailed knowledge of their everyday realities, and there is no hard and fast determinant of the truth, but rather methods and debates for people to achieve some clarity about the world together (Freire, 1981; Greenwood and Levin, 1998, pp85–86). Freire, however, developed his reasoning from the work of Marx and other Marxist thinkers, rather than from pragmatist philosophers. Indeed, Bud Hall (who has been influential in the development of participatory research) suggests the deep influence of Marx in participatory approaches to research in adult education (Hall, 1992). He proposes that PR's origins lie in the research techniques that Friedrich Engles and Karl Marx used in their studies of work conditions in factories during 19th-century France and England.

Freire's approach is firmly anchored in such traditions of emancipating workers, the poor and other oppressed peoples. He argued that ending oppression starts with changing the traditional approach to education. In standard education practice, students are viewed as blank slates that need to be filled up with knowledge which is provided by the teacher. This approach, Freire suggested, 'brainwashes' students into accepting the current system as well as their own oppression. He proposed an alternative approach, embodied in the principle of 'conscientization', in which peasants and students are assumed to come to education already endowed with certain types of knowledge. The teacher does not fill the students with knowledge, but rather poses questions and introduces material to facilitate the process of students and teachers reflecting together on themselves and the world. The goal of education is for teachers and students to work together to arrive at a synthesis of knowledge in which the roots, processes and techniques of oppression are exposed (Freire, 1981).

Lewin and Freire have shaped participatory approaches in city and regional planning and public health as well. With its long tradition of designing cities and landscapes to encourage positive social relations and promote public health, planning practice overlaps with natural resource management. Indeed, as Elmendorf and Rios point out in their contribution to this volume (see Chapter 4), the interaction between the physical and social environments plays an important role in community development and well-being.

Some planners, like their counterparts in conservation and economic development, have argued that the traditional and technocratic top-down approach to designing and planning the physical environment has perpetuated economic inequities and racial segregation. They too have called for a more cooperative approach based on mutual learning between professionals and the public (Warren, 1977; Sandercock, 1998). In crafting their critiques, planners often drew on many of the same sources that have shaped participatory approaches to research and development in CBNRM, including Lewin, Freire, Marx, Dewey and others

(Friedman, 1987; Schön, 1995; Sandercock, 1998; Sanoff, 2000). It is no surprise, then, that participatory approaches to research in natural resources often draw on planning theory and practice.

The same goes for public health. Health workers began to explore participatory approaches to improving health conditions during the 1970s. As in other fields, proponents of participatory research in health argued that health was more a function of people acting and living within their social context than it was of the healthcare system. Participatory approaches to health were therefore intended to improve health overall by reducing dependency upon health professionals, making healthcare programs more sensitive to the social contexts in which people live and ensuring that change efforts have lasting effects (Minkler and Wallerstein, 2003).

In recent years, PR practitioners in CBNRM have begun to tap into this wellspring of experience and insight. In addition, feminist and post-Marxist analyses of the legacy of colonialism, modern nation states and the exercise of authoritarian power have also influenced the thought and practice of some PR practitioners in all of these fields.

Despite the borrowing among different fields and traditions, there are differences in participatory approaches. Although they sometimes overlap, the Freirian and Lewinian trajectories in PR are parallel, rather than interconnected. As Wallerstein and Duran (2003) point out, the key difference between these two major approaches lies in their political projects. Referring to Lewinian approaches as the Northern tradition (due to their origins in European/industrialized settings) and Freirian approaches as the Southern tradition (due to their origins in South American/developing country settings), they observe that the former have more of a problem-solving or utilitarian focus, while the latter focus on emancipation of the poor, minorities, workers, women and other oppressed people (see Table 1.2).

While participatory approaches in the Lewinian tradition are also aimed at bringing about more egalitarian and democratic social arrangements, they do not go quite as far in their critique of global, national and regional political economic systems as approaches in the Freirian tradition.

A useful way of characterizing this difference may be to suggest that on the continuum of participatory approaches, there are those with a radical social agenda and those with more of a social reform agenda. Radical approaches often entail a fundamental questioning of the structure of the global capitalist system and emphasize empowering the poor and marginalized through participatory research (Hall, 1992; Sandercock, 1998). Reform approaches, on the other hand, focus on social change within organizations without questioning the structure of the economic system in which those organizations are embedded, particularly when applied to developed world industrial settings (Whyte, 1991). While such approaches may also be aimed at developing more democratic social arrangements in communities and even fundamental changes in the distribution of power (Greenwood and Levin, 1998), placing the practice of PR in any given case (as well

Table 1.2 *Northern and Southern traditions in participatory research*

Tradition	*Northern*	*Southern*
Assumptions	• Knowledge produced through action • No direct access to objective reality	• Knowledge produced through action • No direct access to objective reality
Goals	• Mutual learning • Social change within organizations or communities • Broader participation in political processes • Better account of the world	• Mutual learning • Freeing marginalized people from conditions of oppression • Fundamental redistribution of political and economic power • Better account of the world
Methods	• Action research	• Conscientization • Participatory action research (PAR)
Origins	• Kurt Lewin • Pragmatism	• Paulo Freire • Marxism

as in any of the fields listed in Table 1.1) on the continuum between utilitarian and emancipatory approaches depends upon the local context, history and ideology of the people involved (Wallerstein and Duran, 2003).

WHAT IS PARTICIPATORY RESEARCH?

Given that there are at least two broad traditions in participatory research, the Northern and the Southern, and that there are many philosophical traditions upon which PR practitioners draw for inspiration and guidance, what do the two traditions and different approaches share in common? Turning to how scholars in different fields have defined participatory research reveals three characteristics that all approaches to PR share. They all entail the production of knowledge through some formal process, they all involve the participation of non-scientists in research processes, and they all are concerned with social change.

Greenwood and Levin (1998, p4), for example, define action research as:

> *... social research carried out by a team encompassing a professional action researcher and members of an organization or community seeking to improve their situation. AR promotes broad participation in the research process and supports action leading to a more just or satisfying situation for the stakeholders.*

Writing in the field of health, Wallerstein and Duran (2003, p28) similarly observe that:

> *... like participatory action research and action research, [community-based participatory research] takes the perspective that 'participatory' research involves three interconnected goals: research, action and education. As part of collaborative democratic processes, shared principles include a negotiation of information and capacities in both directions: researchers transferring tools for community members to analyze conditions and make informed decisions on actions to improve their lives, and community members transferring their expert content and meaning to researchers in the pursuit of mutual knowledge and application of the knowledge to their communities.*

In the field of rural development Robert Chambers (1994, p953) explains that 'PRA [participatory rural appraisal] has been called "an approach and methods for learning about rural life and conditions from, with and by rural people."... The phenomenon described is, though, more than just learning. It is a process which extends into analysis planning and action.' PR thus entails involving the people directly affected by the phenomenon under study in the research process in order to produce new knowledge that can help them to effect social change.

This is simple enough to say. But the involvement of non-scientists as co-researchers in the process of enquiry stems from a fundamentally new understanding of how we can learn about the world. In addition, the goal of effecting social change requires conscious engagement with relationships of power.

Let us look at our understanding of how we can learn about the world – our epistemology – first. Traditionally, scientists have assumed that there is a reality independent of human thought about which scientists, through rigor of method, can uncover the truth. Conventional science rests on the assumption that only trained scientists can produce legitimate findings with a high enough degree of certainty that they accurately portray this independent reality.

In contrast, PR is grounded in the assumption that while there may be a physical reality that exists independently of human thought, knowledge of it is always filtered through cultural lenses. Kurt Lewin, Paulo Freire and the other thinkers in the fields discussed earlier all argued that we do not have access to an objective truth about the world and the things that happen in it; rather, we can engage in dialogues with one another and together develop understandings of how the world and the people and things in it function.

More recently, social scientists and others have further developed this critique of the traditional scientific epistemology. They have specifically questioned the notion that scientists are objective observers who can set aside their biases and remain neutral in the course of research. Daston (1999), for example, has demonstrated that understandings of what constitutes objectivity are themselves produced by the particular scientific ideals, practices and needs of any historical moment. Haraway (1991) has similarly argued that all knowledge is 'situated' within certain historical and social contexts that deeply shape what is deemed the truth.

While such critiques of the traditional scientific epistemology and the traditional practice of science may take extreme forms, there are middle-ground positions, and biophysical scientists are themselves beginning to accept that the current notion of objectivity has been compromised. Allen et al (2001), for example, urge scientists to consider the position that human interaction with the world, through which knowledge is developed, is indelibly shaped by definitions and values. They assert that 'the argument about truth turns not directly on data *per se*, but on the belief that the perception of data yields truth. There can be no such thing as an observer-free observation' (Allen et al, 2001, p475).

Although we cannot directly access the truth, no matter how hard we try to maintain objectivity or how rigorous our methods, we *can* engage in dialogues with one another. Whether one adopts a participatory or conventional approach to research, the outcome of the dialogues in which one engages in the research process are collective social judgments (Greenwood and Levin, 1998) about the situation under study. That is to say, instead of explanations that constitute unchanging, hard and fast truths about objective reality, science produces understandings that the researchers (whether the researchers are all scientists, or a group of collaborating scientists and non-scientists in the case of PR) agree are robust explanations of the situation under study given the state of current theory and knowledge, and the data at hand.

Greenwood and Levin (1998) describe how science is an eminently social activity that produces such collective social judgments. They relate the story of a chemist who lectured about the practice of science to one of Greenwood's undergraduate classes. The chemist explained that scientists work together to develop hypotheses, and stop hypothesis development when they cannot think of any more plausible explanations or are too tired to go on. In other words, their hypothesis generation and subsequent analysis is limited by the extent of their collective knowledge at that particular moment in time, their understanding of current theory, and the nature of the problem they are trying to solve.

Participatory research is every bit as much a social activity. It differs from conventional science in that researchers who adopt a participatory approach explicitly acknowledge that the knowledge science produces is negotiated. Moreover, they seek to expand the pool of people involved in that negotiation – that is to say, the pool of people who engage in the exploration of alternative explanations of the situation under study. The people they seek to include in the dialogue, besides other scientists, are the people who are directly involved with and/or affected by the situation under study. They try to include such people for two reasons:

1 to produce better explanations by incorporating within analyses the knowledge that non-scientists possess about the phenomena with which they are directly engaged; and
2 to address relationships of power that inhibit amelioration of very real human problems such as poverty, income disparity, environmental degradation and conflict over land-use and natural resource management.

People have intimate knowledge of the things with which they are engaged as they go about their daily lives. That knowledge is produced through action. Knowledge of processes and practices (know-how) is tied to the act of engaging in those processes and practices, and is difficult to convey outside of the context of doing them. A brush harvester's know-how, for example, is in the assessments he makes of a patch of brush he encounters in the field, the particulars of the permit that allow him to harvest and, finally, the techniques that he employs in the harvesting itself.

Following applied anthropology, PR recognizes the validity of this knowledge in action as well as the systems of knowledge in which it is embedded, often referred to as local knowledge and indigenous knowledge.[3] Incorporating local and indigenous knowledge and knowledge in action within research enriches the findings because they are an integral part of the functioning of real world phenomena. To omit that knowledge is to exclude a key aspect of how that phenomenon functions (Schön, 1995).

Recognizing the importance of this knowledge and involving the knowledge holders directly in the research in no way constitutes an abandonment of the traditional concerns of conventional science for rigor. Indeed, PR is founded upon disciplined listening and observation and the search for alternate explanations. Like conventional science, it is concerned with the degree to which observations are logically connected to an explanation, and the degree to which that explanation can be accepted as an accurate account of the situation under study. Observation, experimentation and hypothesis testing play the same role in PR as in conventional science; but PR asks questions of them, about power relations and about objectivity, that conventional science does not.

In challenging the notions of scientific objectivity, neutrality and distance from the research subject that are implicit in conventional science, PR calls for rigorous examination of the data collected as well as the research process. PR searches for a negotiated settlement – an agreement among the scientists and non-scientists who collaborate in the research – on both the meaning of the data and the research process itself. Although the research process is often thought of as data collection and analysis, it also entails making choices about the research question, research objectives, data collection techniques, methods of analysis and dissemination of the results. PR uses conventional research tools, as well as tools developed specifically to be participatory, but explicitly sets them in the context of the search for a collective social judgment.

This has implications for evaluating research results. Many PR practitioners prefer the term 'trustworthiness' to 'validity' because the latter implies that the findings of scientific research directly portray objective reality. Validity also implies that research findings can be generalized to fit other cases and broader geographic areas. Trustworthiness, on the other hand, embodies the notion that although we do not have direct access to objective reality, through dialogue and the rigorous exploration of alternative explanations we can produce knowledge that we can trust in making decisions that affect the lives and livelihoods of ordinary people. Depending upon the research methods used, the findings may or may not be generally applied. Assessing the trustworthiness of a study's findings is inherent in PR's self-reflexive nature. PR raises and discusses the effects on the research of blurring the distinction between researcher and 'subject', and of entering the research with an interest in the outcome. This questioning is part of the search for alternate explanations of the problem under study, which, in turn, is crucial to producing well-founded and accurate analyses.

Making decisions that affect the lives and livelihoods of ordinary people brings us back to the second reason that PR practitioners seek to expand the pool of people involved in developing collective social judgments about the situation under study. With roots in concerns for creating more strongly democratic institutions, as well as for addressing the failures of top-down technocratic approaches to economic development and conservation, the practice of PR is aimed at putting science at the service of ordinary people. In addition to producing robust accounts of the world, therefore, PR is also directed at building the capacity of the co-researchers (i.e. the ordinary people) to utilize the research results to change their situation for the better. While government officials, non-governmental organization (NGO) staff, and industrialists are generally positioned economically, socially and politically to have ready access to the findings of scientific research, as well as to the resources they need to utilize those findings, ordinary people – especially the poor, racial and ethnic minorities, many women and other marginalized people – do not.

Researchers benefit from research by taking the information they extract from communities and publishing it, lecturing about it or otherwise applying it in ways that advance their own careers. The information and research results may be used by other entities external to the community to their own benefit as well. Often the research results are irrelevant to the communities and of little use for solving the problems that they face. In some cases the community may become even worse off as a result of the research. For example, Starn (1986) has demonstrated how the well-intentioned research of anthropologists on the War Relocation Authority camps during World War II reinforced stereotypes of Japanese Americans, limited public debate and legitimated relocation. Although addressing this issue of extractive research is important to many PR practitioners, the practice still continues even under the name of PR (Simpson, 2000). Extractive research affects both indigenous and non-indigenous communities; but native peoples began addressing it proactively and attempting to control research projects conducted in/on/with them more than three decades ago after Vine Deloria Jr published *Custer Died for Your Sins* (Biolsi and Zimmerman, 1997; see also Chapter 11 in this volume).

The new epistemology bears directly on this issue. Acknowledging that all knowledge is situated within specific historic and social contexts invites respect for other knowledge systems. This opens the door to ordinary people being positioned to contribute to, and directly benefit from, research, as well as to use research results themselves to improve their own lives or have a voice in their own affairs. That is to say, the new epistemology is crucial to preventing extractive research and to rebalancing the relationships of power in traditional research that tended to benefit people outside of the community.

Addressing power relations is often put into practice through building the capacity of community members to more actively determine their own futures. In PR, the research process is as important as the research findings because it is through that process that capacity-building is thought to occur. The goal is for community members to develop research skills as well as the competency to use those skills to address their own problems. As they identify the research questions and carry out research activities, community members learn to analyze information they have collected and decide how to use this information. Most important, communities 'own' their research. That is to say, they have intimate knowledge of the research procedures and findings, and feel comfortable using or disseminating those findings themselves. Depending upon the specifics of the project, as well as local circumstances, the research process is thus intended to contribute to enhancing the capacity of community members to do better any or all of the following: mediate their own conflicts, represent their interests in wider social and political arenas, manage the resource sustainably, participate as informed actors in markets, build community assets with benefits from managing the resource (Menzies, 2003), and sustain their own cultures.

Achieving these and other goals of PR is not straightforward, as recent critiques of participatory processes have made clear (Guijt and Shah, 1998; Cooke and Kothari, 2001; Lane and McDonald, 2005). The practice of PR requires making choices throughout the research process that affect the research results, as well as the building of capacity among community members. Choices must be made in all research about which field methods are most appropriate to use, what variables require the closest scrutiny, and how to design the research to most effectively answer the question being asked. Accepting the notion that we do not have direct access to objective reality, but, rather, that we can learn together through dialogue, entails choices in addition to those usually encountered in research. Decisions about who participates in the dialogue, for example, affects research objectives as well as what questions will be asked, and, therefore, what issues the findings will address. If only scientists are included in the research process, the tendency will be for just those questions of interest to scientists to be investigated. This, of course, lies at the root of the problem of scientific studies, often having little direct relevance to communities. PR practitioners must therefore make choices in every PR project about who participates, how and at what stages.

Even the extent to which relationships of power are critically examined is a choice. Although it is key to effecting social change, the degree to which co-researchers wish to address uneven relationships of power, within communities as well as between communities and external entities, varies from project to project, and may vary among the co-researchers in a single project. Likewise, PR projects vary in the extent to which they are directed at building capacity among community members. Some proponents of PR, such as Minkler and Wallerstein (2003, p7), assert that PR projects with emancipatory goals should be the highest standard of participatory practice. Hall (1992) similarly suggests that PR must benefit marginalized people (the poor, the oppressed and the disenfranchised). Although other proponents of participatory approaches, such as Greenwood and Levin (1998) and Whyte (1991), are concerned with creating more democratic structures and processes, they do not specifically mention marginalized people as a particular target of participatory approaches.

OUTLINE OF THE BOOK

There are many issues central to the current practice of PR that require making choices. There are no easy answers to the questions these issues raise, and there are no templates to apply in every situation. Rather, these issues necessitate choices among alternative actions that must be negotiated anew among the co-researchers in every PR project. The authors of the chapters of this book discuss how they dealt with these ongoing issues and the implications they had for empowering the community as well as for the research results. Their chapters show that PR can produce the benefits that its proponents claim, but that this takes negotiation and commitment, and that research design must be tailored to local context.

Tailoring research to the local context raises the question of the degree to which it is necessary for PR projects to share features and characteristics. This question is addressed in Chapter 2. Observing that there are many different approaches to participatory research, the authors distinguish between participatory action research and participatory research, and suggest that there are three criteria essential to the success of these approaches:

1 the degree of community-centered control;
2 the reciprocal production of knowledge; and
3 the utility and action of outcomes.

They suggest that these criteria constitute key elements of PAR and other participatory research approaches, and can be used for evaluating individual projects as well as the usefulness of the larger methodological approach. They caution, however, that any particular description of these criteria, including their own, should not be considered to have universal applicability. Because every situation in which research is conducted is different, the criteria need to be flexible to accommodate the particulars of each case.

While these criteria may serve as useful guides to conducting research that empowers communities, broadening the scope of participatory research such that it has greater impacts in increasing numbers of communities depends upon its acceptance within the broader educational and community development systems. Chapter 3 examines challenges to the institutionalization of participatory research through a comparison of an effort in the academy and an effort in a non-profit organization to institutionalize it. The authors argue that the separation of research and community development within the current educational system leads to career incentives, structural barriers in education, funding priorities and funder requirements that inhibit the institutionalization of participatory research, which, in turn, hinders community empowerment through the practice of PR.

Following Chapters 2 and 3 and their analysis of more general issues, Chapters 4 to 7 analyze issues in the practice of community capacity-building through participatory research. Chapter 4 discusses community development as relationship-building in the context of a community-based environmental improvement project in what is considered the most socio-economically disadvantaged neighborhood in Philadelphia, Pennsylvania. The case counters the stereotype of African Americans having little interest in the environment, and shows how collaborative planning and joint construction of urban parks and gardens builds trust and interaction between local residents and local and outside organizations. This interaction also enables community residents to take on more complex projects, which the authors argue is the essence of community capacity-building.

Empowering communities may also involve creating opportunities for community members to have a greater voice in managing whole watersheds as is demonstrated in Chapter 5. The chapter describes how an undergraduate student's senior project in landscape architecture grew organically into a community-based effort to restore a native California riparian ecosystem on the site of a former gravel quarry. The authors demonstrate how a participatory approach to research has facilitated collaboration between local landowners, county officials, the gravel industry and local Native Americans, and enabled the latter to become players in the management of the local watershed in a manner that had previously eluded them. Achieving this level of community empowerment was not easy, however, and the authors discuss why, paying special attention to issues that the various groups involved encountered in establishing trust with one another.

Chapter 6 raises questions about how local power relationships affect the effectiveness of participatory research in serving the communities in which it takes place. The authors draw upon their involvement with a participatory research initiative in Macon County, North Carolina, where second-home development is rapidly transforming the Southern Appalachian Mountains. By fostering meaningful public dialogue on values and land use, the project aimed to help build the community's capacity to determine the future of their local landscape. Visions articulated by project participants, however, have not readily translated into long-term empowerment or altered land-use outcomes; therefore, community members and researchers are now undertaking an evaluation process aimed at broadening and sustaining the initial dialogue. The authors discuss both the strengths and the limitations of participatory research in effecting meaningful community empowerment, and they propose research design and evaluation strategies that can improve the chances for success.

Additional insights about the obstacles to empowering communities are offered in Chapter 7, which examines in detail the collaborative process of revising the forest plan for the Grand Mesa, Uncompahgre and Gunnison National Forests. Through their frank and honest discussion of the many personal discomforts, procedural obstacles and differences of opinion encountered in the process, the authors illustrate how PR and the collaborative management of natural resources truly are fluid and contested processes. They conclude with a discussion of several lessons learned in the process, including the need to build systematic monitoring approaches to collaborative processes at the very beginning of a project, clearly defining roles and responsibilities, and continuously working to ensure open and honest communication among all parties involved.

Chapters 8 and 9 should be read as a pair. Each chapter compares two case studies to draw lessons for conducting participatory research, and the case of a research project with salal harvesters on the Olympic Peninsula in the state of Washington is common to both chapters. The chapters ask different questions of the case studies, however. In Chapter 8, the authors ask how participatory research can be used to overcome the exclusion of marginalized social groups (in this case immigrant Latino forest workers) from the science, policy and management of forests and other resources. They identify several factors that contribute to the exclusion of these workers from the civic life of the communities in which they work, including language barriers, undocumented immigrant status, intra-ethnic exploitation and hierarchical labor relations, racism and other structural factors. They note that these same factors turn out to be strong inhibitors of the involvement of forest workers in participatory research. Overcoming these barriers meant working through trusted, established non-profit organizations that provide various social services to the workers.

Chapter 9 compares the Latino salal harvesters case study with a participatory research project on harvesting salal that Heidi Ballard conducted with the Makah Indian tribe in Washington. The question that guides the discussion in this chapter is how can participatory research contribute to ecological science? The comparison demonstrates the importance of context to the success of participatory research. While the Latino harvesters contributed valuable knowledge of the resource to the study, knowledge which added significantly to the quality of the research, several aspects of how the Makah study was conceived and implemented led to less participation and fewer significant contributions by tribal members. The authors conclude that continuity of involvement by at least some of the same community members is crucial to achieving the benefits of participatory research, and that the presence of a strong leader with natural resource management and human resource management skills is necessary. They also suggest that characteristics of communities, such as cohesiveness and long-term presence in a place that are often assumed to lead to receptivity to participatory research, may actually serve as obstacles in some cases.

This points to the need for skills in translating between the distinct worldviews of different cultures in research. This need is discussed in Chapter 10 as it arose in response to an incident on the White Mountain Apache Reservation in Arizona in which federal land managers wanted to clear debris from a stream in a manner at odds with the wishes of many community members. The problem was more than just needing to have the community's voice heard in the management decision. What was necessary were individuals who could translate knowledge between community members and outsiders effectively so that a more comprehensive understanding of the problem could be achieved. The authors discuss four 'pathways' to sharing knowledge between different management traditions, and suggest that participatory research can facilitate such sharing because of its focus on the tensions between insider and outsider knowledge.

Chapter 11 takes the issue of translating across cultures a step further by discussing the dilemmas that Native American scholars face. These dilemmas stem from stereotypical perceptions of Native Americans and native communities that effectively recapitulate the past exclusion of Native American communities from research and lead to extractive research. The authors discuss strategies for overcoming these dilemmas, including developing relationships of reciprocity and improving communication across cultures.

NOTES

1 For a definition of CBNRM, see Box 1.1.
2 RRA entails short visits (often about two weeks in duration) to rural sites by a team of researchers who observe livelihood practices, interview residents, map land-use patterns, hold group discussions, and use other data-gathering methods. The intent of this chapter is not to provide a guide to field techniques. The interested reader may wish to consult works by McCracken et al (1988), Slocum et al (1995) or Gonsalves et al (2005).
3 Indigenous knowledge is knowledge produced by the knowledge systems of native peoples. Local knowledge, in contrast, refers to non-scientific knowledge produced by non-indigenous people living in specific localities. Depending upon the circumstances, indigenous, local and scientific knowledge may overlap and shape one another. A good example of the interdependence of scientific and local knowledge is Edward Jenner's invention of the smallpox vaccination. Farmers in western England where he worked told Jenner that people who developed cowpox from milking cows usually did not develop smallpox. Jenner reasoned that in fighting off cowpox, a less deadly relative of smallpox, the body somehow developed the ability to ward off smallpox. He further reasoned that inoculating people with cowpox would render them immune to smallpox. With the assistance of local knowledge, he had discovered vaccination.

References

Allen, T. F. H., Tainter, J. A., Pires, J. C. and Hoekstra, T. W. (2001) 'Dragnet ecology – "Just the facts, ma'am": The privilege of science in a postmodern world', *BioScience*, vol 51, no 6, pp475–485

Baker, M. and Kusel, J. (2003) *Community Forestry in the United States: Learning from the Past, Crafting the Future*, Island Press, Washington, DC

Biolsi, T. and Zimmerman, L. J. (eds) (1997) *Indians and Anthropologists: Vine Deloria Jr and the Critique of Anthropology*, University of Arizona Press, Tucson, AZ

Chambers, R. (1980) 'Rapid Rural Appraisal: Rationale and Repertoire', IDS Discussion Paper No. 155, Brighton

Chambers, R. (1994) 'The origins and practice of participatory rural appraisal', *World Development*, vol 22, no 7, pp953–969

Chambers, R. (2005) *Ideas for Development*, Earthscan, London

Cooke, B. and Kothari, U. (eds) (2001) *Participation: The New Tyranny?*, Zed Books, London

Cornwall, A. and Jewkes, R. (1995) 'What is participatory research?', *Social Science and Medicine*, vol 41, no 12, pp1667–1676

Daston, L. (1999) 'Objectivity and the escape from perspective', in M. Biagioli (ed) *The Science Studies Reader*, Routledge, New York, NY

Flower, C., Mincher, P. and Rimkus, S. (2000) 'Overview – Participatory Processes in the North', PLA Notes, no 38, pp14–18

Forester, J. (1989) *Planning in the Face of Power*, University of California Press, Berkeley, CA

Freire, P. (1981) *Pedagogy of the Oppressed*, Continuum, New York, NY

Friedman, J. (1973) *Retracking America*, Doubleday Anchor, New York, NY

Friedman, J. (1987) *Planning in the Public Domain: From Knowledge to Action*, Princeton University Press, Princeton, NJ

Gaventa, J. (1980) *Power and Powerlessness: Quiescence and Rebellion in an Appalachian Valley*, University of Illinois Press, Urbana, IL

Gonsalves, J., Becker, T., Braun, A., Campilan, D., De Chavez, H., Fajber, E., Kapiriri, M., Rivaca-Caminade, J. and Vernooy, R. (eds) (2005) *Participatory Research and Development for Sustainable Agriculture and Natural Resource Management: A Sourcebook*, International Potato Center-Users' Perspectives with Agricultural Research and Development and International Development Research Centre, Laguna, The Philippines, and Ottawa, Canada

Goodman, R. (1972) *After the Planners*, Penguin, London

Gray, G. J., Enzer, M. J. and Kusel, J. (2001) 'Understanding community-based forest ecosystem management: An editorial synthesis', *Journal of Sustainable Forestry*, vol 12, no 3/4, pp1–23

Greenwood, D. J. and Levin, M. (1998) *Introduction to Action Research: Social Research for Social Change, Sage*, Thousand Oaks, CA

Gujit, I. and Shah, M. K. (eds) (1998) The Myth of Community: Gender Issues in Participatory Development, Vistaar Publications, New Delhi

Hall, B. (1992) 'From margins to the center? The development and purpose of participatory research', *American Sociologist*, vol 23, no 4, pp15–28

Haraway, D. (1991) 'Situated knowledges: The science question in feminism and the privilege of partial perspective', in D. Haraway (ed) *Simians, Cyborgs, and Women: The Reinvention of Nature*, Routledge, New York, NY

Lane, M. B. and McDonald, G. (2005) 'Community-based environmental planning: Operational dilemmas, planning principles and possible remedies', *Journal of Environmental Planning and Management*, vol 48, no 5, pp709–731

Lewin, K. (1948) *Resolving Social Conflicts*, Harper, New York

McCracken, J. A., Pretty, J. N. and Conway, G. R. (1988) *An Introduction to Rapid Rural Appraisal for Agricultural Development*, IIED, London

Menzies, N. (2003) Seminar presented at the Division of Society and the Environment, University of California, Berkeley, CA

Minkler, M. and Wallerstein, N. (eds) (2003) *Community-based Participatory Research for Health*, Jossey-Bass, San Francisco, CA

Murphree, M. W. (1993) 'The Role of Institutions', paper presented at Workshop on Community-based Conservation, Washington, DC 15–22 October

Sandercock, L. (1998) *Toward Cosmopolis: Planning for Multicultural Cities*, John Wiley and Sons, Chichester

Sanoff, H. (2000) Community Participation Methods in Design and Planning, John Wiley and Sons, New York, NY

Schafft, K. A. and Greenwood D. J. (2002) 'The Promise and Dilemmas of Participation: Action Research, Search Conference Methodology and Community Development', Paper submitted to the 65th Annual Meeting of the Rural Sociological Society, Chicago, IL

Schön, D. A. (1995) 'Knowing-in-action: The new scholarship requires a new epistemology', *Change*, vol November/December, pp27–34

Simpson, L. (2000) 'Aboriginal peoples and knowledge: Decolonizing our processes', *The Canadian Journal of Native Studies*, vol XXI, no 2, pp137–148

Slocum, R., Wichart, L., Rocheleau, D. and Thomas-Slayter, B. (eds) (1995) *Power, Process and Participation: Tools for Change*, Intermediate Technology Publications, London

Starn, O. (1986) 'Engineering internment: Anthropologists and the war relocation authority', *American Ethnologist*, vol 13, pp700–720

Wallerstein, N. and Duran, B. (2003) 'The conceptual, historical, and practice roots of community based participatory research and related participatory traditions', in M. Minkler and N. Wallerstein (eds) *Community Based Participatory Research for Health*, Jossey-Bass, San Francisco, CA

Warren, R. L. (1977) *New Perspectives on the American Community: A Book of Readings*, Rand McNally, Chicago, IL

Western, D. and Wright, R. M. (1994) 'The background to community-based conservation', in D. Western, R. M. Wright and S. C. Strum (eds) *Natural Connections: Perspectives in Community-based Conservation*, Island Press, Washington, DC

Whyte, W. F. (ed) (1991) *Participatory Action Research*, Sage, Newbury Park, CA

Wilkinson, K. P. (1991) *The Community in Rural America*, Greenwood Press, Westport, CT

2

Core Criteria and Assessment of Participatory Research

J. D. Wulfhorst, Brian W. Eisenhauer, Stephanie L. Gripne and Johanna M. Ward

COMMUNITY-CENTERED PARTICIPATORY RESEARCH

Conducting research that affects constituencies of interest while producing general knowledge has long been a challenge for application-oriented researchers. Different approaches have emerged to achieve that elusive goal. Some of these approaches emphasize the necessity of working in collaboration with communities. However, considerable debate about the optimal way to conceptualize and conduct collaborative research activities as community-based approaches continues (Powers et al, 2006).

As a category of approaches, community-centered participatory research (PR) is research by and with a community rather than simply for or about a community. PR offers an alternative approach to learning within scientific enquiry because it responds directly to community needs by incorporating conceptual and methodological ideas and direction from community representatives participating throughout the research process, instead of treating participants simply as subjects of, or bystanders to, the research process (Greenwood and Levin, 1998b). PR intentionally generates community-based benefits by focusing on process, inclusion and application. Therefore, PR fosters representation and solutions that draw from the perspectives of participating individuals and group representatives affected by potential actions generated within research (Pretty, 1995).

Generally, PR moves beyond traditional approaches to research by virtue of its intention to not only generate valid knowledge, but to more directly affect the communities involved through their participation in all stages of the research process (design and conceptualization, data collection, analysis and reporting).

This model expands scientific enquiry to include the goal of local empowerment and building community capacity through the facilitation and application of local knowledge in the research process (Sillitoe, 2006).

More specifically, participatory action research (PAR) exists along the PR continuum as an important set of theoretical and methodological considerations. Instead of relying on outside experts to write a prescription for a community, PAR facilitates community control of research processes in the communities' own efforts to affect change. This methodology has the potential to empower because community members, rather than experts, actively lead and control aspects of projects in order to build community capacity and influence the future of their communities. Projects using this methodological approach may reflect and incorporate community sentiments and needs more than those guided by more traditional expert-driven research models. As summarized by one research team, community members 'did not want social agents to come and solve their problems, but resources in order to do it themselves. They want[ed] to be the protagonists of their own change' (Crespo et al, 2002, p54).

But what are the criteria that define and distinguish participatory approaches? Over a decade ago, Pretty (1995, p1251) already noted that PR 'is such a fashion that almost everyone says that participation is part of their work'. So how do we identify what constitutes PR and what does not? Is there a continuum on which we can place or evaluate any project to say one adheres to PR and another does not? Perhaps more importantly, if the distinctions between PR and other methodological approaches are relevant, they must be clarified before a meaningful and systematic evaluation of PR as a whole, or a specific PR project, is possible. Implicit in this assertion is the notion that PR is equally credible and relevant to on-the-ground realities as traditional positivist approaches and programs (Greenwood and Levin, 1998b).

The lack of clarity about core criteria within the overall field of participatory research presents challenges of consistency to new and experienced practitioners alike attempting to apply these strategies in the field. The diversity of perspectives and community realities, however, also begs the question of whether consistency is appropriate and pertinent to PR approaches. Because PR emphasizes community engagement in the design of research and the need to tailor research projects, it is reasonable to ask what features and characteristics PR projects should share. Thus, there are important questions to be examined in order to better understand and improve PR: are there standardized criteria? If so, how do researchers integrate their academic goals with the emphasis on building community capacity and the opportunity for local empowerment? Together, these questions explain Cornwall and Jewkes' point (1995, p1667): 'What is distinctive about participatory research is not the methods, but the methodological contexts of their application.' Similarly, Pain and Francis (2003, p46) noted: 'the defining characteristic of participatory research is not so much the methods and techniques employed, but the degree of engagement of participants within and beyond the research encounter'.

In this chapter, we focus on whether PAR constitutes a unique form within the range of PR approaches, as some practitioners claim, in light of the criteria we propose, with an eye to answering the above questions. To this end, we review the core themes of PAR to emphasize its explicit dual intent to act as both a research approach, as well as a transformational platform for interested communities, groups or organizations. The chapter then describes the three criteria essential for the success of these approaches:

1 community-centered control;
2 reciprocal production of knowledge; and
3 action outcomes and who benefits.

Not intended as comprehensive or exclusive, we argue that these criteria embody core principles and fundamental measures to consider when designing PAR efforts where issues of community participation and change are critical.

PARTICIPATORY RESEARCH AND PARTICIPATORY ACTION RESEARCH

Very generally, previous literature reveals that participatory research is referred to as a paradigm, method (McTaggart, 1991; Finn, 1994; Guevara, 1996), framework (Guevara, 1996), approach, set of tools (Sims and Bentley, 2002), model (Guevara, 1996; Sims and Bentley, 2002), research strategy (Greenwood and Levin, 1998a, 1998b; Guevara, 1996), worldview (Reason and Bradbury, 2001) and consulting technique (Sims and Bentley, 2002), despite multiple calls for greater clarity by practitioners (Conchelos and Kassam, 1981; Vio Grossi, 1981; Schroeder, 1997).

Participatory research presumes that people ought to, and will, engage in civic action to address issues that affect the quality of their lives. PR operates similarly to at least some collaborative resource management approaches, such as grassroots ecosystem management (GREM) (Weber, 2000), by virtue of its focus on designing and conducting research as a consensual process. PR projects, however, extend beyond the goal of bringing community interests together to make decisions. PR projects engage local citizens in the design of scientific processes to ask researchable questions and to take action in the course of developing and maintaining community members' involvement (Cornwall and Jewkes, 1995; Caldwell et al, 2005). As a research model, participatory approaches provide those affected with the opportunity to help guide what issues the research should focus on, assist with the process of defining and articulating the research questions, and facilitate and conduct the investigation with the intent of applying the findings in the community.

Similar to how Pretty (1995) noted the intrinsic problems of defining the concept of 'sustainability', PR defies any singular definition. Greenwood et al (1993) argued that PR exists on a continuum where the range and degree of participation vary. In fact, PR could encompass both very passive and very active types and levels of participation. Because the range of participation described is inclusive, however, such an open-ended approach does not demarcate what is PR and what is not. Maguire (1993) explored the struggles that researchers often face with PR due to the normative expectations associated with standardized criteria within traditional scientific methods (i.e. being an organizer and advocate, in addition to collecting data, theorizing and analyzing with academic frameworks). Similarly, Yung's (2001) research encouraged researchers to incorporate PR practices within their efforts, even when research settings do not lend themselves to every component of a participatory research approach.

Some distinguish PR from PAR, while others consider them one and the same (for discussion of this point, see Park, 1999; Reason and Bradbury, 2001). Here, we highlight PAR within the range of PR approaches in order to emphasize precisely the valuable nuances and perspectives of this range. Amidst the debate on whether we can differentiate PAR from PR (Park, 1999; Reason and Bradbury, 2001), contrasting these ideas is a useful analytical exercise to achieve the goals of this chapter. Both PR and PAR share the idea that in order to conduct meaningful research, community members must not be placed 'under the magnifying glass' of experts from the outside. Within all participatory approaches the validity of the enquiry is based, in part, on the process that facilitates community members' involvement in defining and designing all stages of the research process as active participants.

However, some literature (see Whyte 1991, 1998; David, 2002) claims that PAR does more than just involve subjects in the research process to obtain valid knowledge. PAR also facilitates achieving the goals of the community being researched through participants' involvement:

> *While traditional forms of participatory research rejected detachment for involvement, PAR goes one step further. The researcher does not simply engage; they engage to facilitate the goals of the researched.* (David, 2002, p14)

Critical to successful PAR projects is the recognition that the design of research and the actions resulting from the application of findings are not separate, as is perceived in more traditional models of research (Greenwood and Levin, 1998b). Instead, the core philosophy of PAR is that optimal research and action outcomes occur when these processes inform one another:

> *As a* research *methodology,* participatory action research *makes it possible to search for solutions to real problems by establishing a closer link*

> *between* research *and* action. *Such a link is possible if the community, the traditional* research *subjects, share a more active role together with the professional researcher during the entire* research *process.* (Guevara, 1996, p32)

The research process itself and the changes in the community that result from being involved in it may be as important for communities as the information collected and analyzed (Wismer, 1999; Mordock and Kransy, 2001). Tangible changes can include the acquisition of research skills and empowerment that enables the community to design an approach in order to address future problems. Lamenting the 'decades it has taken the scientific community to recognize that Native communities can identify their needs, determine courses of action and achieve the goals they have set for themselves', Caldwell et al (2005, pp8–9) discuss the specific case of applying PAR in American Indian and Alaskan natives' contexts. They document how the process of 're-traditionalization' lends itself to a methodological approach that is not only participatory, but also addresses a community's needs by tapping into the strengths and capital of the local population.

Clarification of this point elucidates not only the structure of PAR, but also a key assessment factor for PAR – namely, the continuity of participation across all stages of the research process, from conceptualization to design, from data collection to analysis, outcomes and actions. Building community capacity and an end state of democratic community cooperation can present challenges for a variety of reasons. As a social process, PAR involves social negotiations, uncertainties and risks. In any given set of circumstances, conditions or contexts, the effects of these dynamics may enable or prevent PAR as a successful approach.

PAR AND NATURAL RESOURCE MANAGEMENT

Communities once considered natural resource dependent from historical patterns of extraction have, in many cases, experienced poverty, cyclical employment and/or overexploitation of local resources. Often studied through the lens of the outside or expert researcher, investigations of the complexity of these communities' dynamics with regard to identity, livelihood and collective morale resulted in research findings that often failed to meet the needs of local interests, even if benefits were indirectly expected and intended on the part of the researchers (Carr and Wilkinson, 2005). Within the context of natural resource management, participatory action approaches have enabled diverse groups of stakeholders to not only increase input, but, more importantly, to establish a fundamentally different platform to provide local knowledge (Michaels et al, 2001; Purnomo et al, 2004).

The historical roots of PAR in natural resource management can be traced back to development work in non-industrialized countries (Pretty, 1995; Levin, 2003) where the methodology emerged as an alternative to top-down approaches that had often resulted in failed applied projects. From this core, community involvement always stood as a central tenet of PR with the goal of coordinating local knowledge and resources to address the issues that affect local constituencies. Finn (1994, p30) articulated how the inspiring work of Paulo Freire's educational approach provided rationale for PAR through the 'decoding of dominant discourses about social problems, pooling knowledge and experience, questioning conditions that affect people's lives and opening dialogue to generate critical thought and action'.

This perspective, when viewed in light of the assertion that community members must be engaged throughout the research process in PAR, clarifies that the researcher's role involves more than just technical expertise; it also involves acting as a facilitator to engage community members across the dimensions of the research project. Such an undertaking requires trust and it can be challenging for a researcher from outside the community to obtain the levels of trust needed to facilitate PAR. To address this, researchers using participatory action approaches conventionally live in the communities with whom they work. Doing so can support efforts to establish credibility and understand different aspects of how one could approach the community and research questions. The facilitation role is central to PAR. In assuming this role, professional researchers do not disavow their expertise, but instead employ it in conjunction with community members to apply participants' knowledge and input to the research issues.

The development of skills and the transfer of knowledge within and between groups empower people in communities; thus, action-directed research aiming to improve people's lives is PAR's desired outcome (Flora et al, 1997; Gaventa and Cornwell, 2001). PAR makes no claim to either objectivity or value neutrality and rejects researcher control in favor of community empowerment to define appropriate and relevant research questions, as well as methodological choices. In essence, PAR combines the production of knowledge with action in order to promote a collective and more integrated research effort that remains locally grounded, as well as process and outcome oriented (Park, 1993; Finn, 1994).

To generate critical thought and action (Finn, 1994), recent works on the roles of local and scientific forms of knowledge in environmental controversies (Fischer, 2000) cite that techniques such as PAR have the potential to enable the ecological (or reflexive) democracy called for by Beck (1995) and others (Fischer, 2000). These perspectives assert that during the early 21st century social affairs are characterized by the 'risk society', a term coined by Beck (1995) to describe a social reality in which the primary policy and environmental questions revolve around the technical and scientific assessment of risk, such as environmental risk. This state of affairs places the locus of power in the disciplines of scientific expertise (Foucault, 1980; Fischer, 2000), as only technical disciplines have the positivist methods, disciplinary norms and expertise in language to make claims about risks that are considered valid and reliable in a risk society. Power is then located in the disciplines and management agencies, which determine not only what constitutes valid information that can be used in a debate, but also the parameters that define an issue and the way in which it is approached. These conditions remove citizens from most policy debates, furthering the chance of a misinformed public, particularly regarding those debates that concern environmental and resource management issues (Phadke, 2005; Norgrove and Hulme, 2006). And as often found within the dynamics of environmental conflict resolution (Crowfoot and Wondolleck, 1990), even participatory approaches may remain a function of power and authority structures embedded in the institutional arrangements of research (Quaghebeur et al, 2004). Considering overlap or parallels between PAR and other collaborative methods may enable researchers and community members to 'accommodate new forms of knowledge, multiple sources of information and balance both expert and lay input' (Moote et al, 2001, p100).

In this context, the challenge for those seeking to involve communities in determining their own future and for democratic ideals as a whole (Beck, 1995; Fischer, 2000) is to open the disciplinary forums of risk debates to public participation. PAR, in its effort to inclusively and consistently involve citizens, offers a methodology for citizens to engage in these debates and to bring local knowledge and perspectives into the process via a community-centered research platform and context. The opportunity to broaden what is considered relevant knowledge and viable ways of knowing essentially alters the form in which power is currently structured in most natural resource-related controversies. In combination, the unique inclusive methodology of PAR, its potential for critical thinking, its mission to involve citizens in scientific and disciplinary research, and the emergent nature of the research process have the potential to empower citizens and to shape collaborative policy debates. In addition, these dimensions offer many challenges to researchers using PAR approaches within institutions and agencies that typically govern and fund research projects. As such, PAR has great potential for shaping the role of expertise and power in environmental controversies if the above challenges can be met.

Use and application of PAR within many projects and communities of the Community Forestry and Environmental Research Partnerships (CFERP) program has established relationships between researchers and an innovative spectrum of community-based constituents to facilitate sharing experiences, strategies and knowledge. With an emphasis on community forestry, the CFERP has offered a critical mass of projects and personnel. The long-term nature of the program has stimulated broader ongoing discussion and interaction about assessment.

Evaluating PAR begs determination of the scope and purpose of such assessment. This is particularly challenging for PAR projects in natural resource-dependent communities because different levels of analysis exist that imply different approaches to conducting the evaluation. These challenges raise the question of whether PAR methodologies effectively produce useful community information and academic knowledge. Similarly, but on a different scale, there is also a need to assess how well PAR goals are being met within specific projects. The following sections outline and discuss three core criteria that form a basis for considering how to evaluate PAR as its applications broaden in community settings with paramount natural resource issues.

Criterion one: Community-centered control

The first criterion identified for evaluations of PAR efforts – community-centered control – is a fundamental aspect of PAR. Three key principles that guide the notion of PAR as community-centered are essential in community control: community ownership, credibility and continuity of trust. Each of these points contributes to the successful structuring of community control, as well as to an understanding of it as a symbolic process of establishing relationships, authority and decision-making power between researchers and community representatives in each research phase. Concepts of community within rural and natural resources sociology remain problematic and under debate. Thus, defining community control may also be a function of the definition of community that one employs (Wilkinson, 1991); but our intent is to use the term to indicate the set of relations that contribute to building capacity and local development of people and places.

A community-centered project produces short-term and long-term benefits for communities in multiple ways, each importantly tied to the level of control that rests within the community or group that defines the problem, need and/or issues. Community-centered research can ensure that the immediate efforts more directly fit local participants' perceptions and needs. But, if a researcher intends to arrive in a community with time and resources to conduct a study, at what point does she or he invite community members' perspectives about the role of the community and focus of the project? Being community-centered allows local participants to build capacities in the community that enhance the ability to address needs through future collaboration with researchers. However, even when research has a community-centered approach, ongoing determination of who will guide the research, as well as how and when, is dynamic. Researchers must recognize that communities do not exist in monolithic or static states, but rather constitute living entities in which political, economic, cultural and a host of social processes of negotiation naturally recur. Accordingly, engagement in and with communities should occur as early in the research process as possible to maximize the research team's knowledge of, and ability to, work with the wide variety of dynamic factors within a community relevant to a PAR effort.

Community ownership

No one prescription exists as to how a community can or does control a project. However, PAR relies on the formation of community ownership and empowerment, both of which remain functionally tied to involvement throughout the research process that includes responsibility, knowledge acquisition and research skill development. In a research model that builds these capacities, a project must be community-centered – an orientation fundamentally different than simply providing opportunities for input.

Participation constitutes more than community members merely telling researchers what they desire to happen. In PAR, participation also involves community members as researchers themselves. Helping to design data collection instruments, critiquing what analyses may be useful and identifying the techniques that can best accomplish project goals are all parts of the research process that enable community control in PAR. Taking part in these efforts not only ensures that the current project meets community desires and needs, it also builds the capacities of those who remain when the research process is complete (Wilkinson, 1991; Simonson and Burshaw, 1993). Capacity-building characterizes the empowering dynamic in the PAR process in part because these processes may enable, motivate and mobilize communities to address other needs in the future (Eisenhauer and Wulfhorst, 2005). Once the utility of research is recognized and some of the skills needed to design and conduct it are developed, communities may collaborate with researchers in the future and engage in research projects again when the need arises.

The action orientation of PAR is a particularly important factor for community ownership and how the locus of that control differentiates PAR projects from a continuum of more traditional academic approaches, as well as from other participatory approaches. For communities, the motivation to conduct research is to affect their homes, livelihoods and collective sense of well-being through action. In contrast, academics are often more concerned with the production of general knowledge to contribute towards greater understanding.

In their discussion of some of the barriers to enhancing community participation, Baker and Kusel (2003) point out that the focal points often dominating local perspectives may also inhibit the active involvement of those whom PAR intends to engage. The dual purposes of community empowerment and the pursuit of knowledge do not immunize PAR from local politics and other community-based factors that influence the diversity and levels of participation. Thus, researchers should not be surprised if local participants and stakeholders are less concerned with how the research contributes to a body of knowledge in science than whether the information has a direct impact upon, and application to, their lives and communities.

PAR must engage the community as a partner in the ownership and control of decisions, answers and processes through the facilitation of continued community involvement throughout the research process. Collaborative techniques, including focus groups, planning sessions and workshops, are integral parts of these processes, which ensure that the community is continually engaged, not simply present. Researchers must facilitate by going beyond informing community members to enabling them as decision-makers who affect all stages of the research process.

How, then, do the community-centered criteria of PAR fit into assessment? PAR assessments should identify the relationships between the degree of a shared sense of ownership and the prevalence of desired outcomes by examining the results of multiple projects. Applying a PAR methodology in natural resource-dependent community settings has highlighted the opportunity for research insights to stem from local ecological knowledge (Medley and Kalibo, 2007). To put such analyses in the context of PAR criteria, we need to consider community members' participation, understanding and ability to make use of outcomes. However, not all PAR projects result in outcomes deemed successful or satisfactory to all participants (Powers et al, 2006).

Thus, within a PAR project, another important evaluation component is the assessment of community members' perceptions of their involvement in the process. As part of a formative evaluation that can improve a PAR project in process, periodic queries of community members about their perceptions of, and satisfaction with, their involvement and degree of ownership within the project (Bailey, 1992) provide insight as to whether the community-centered criteria are being met. Although these suggestions identify only the very first steps of conducting such evaluations, clarifying the need to consider the importance of a sense of ownership within a community-centered orientation is an important component of determining success.

Credibility

Also essential to the idea of a project being community-centered is the sense of credibility that must exist between participants and researchers. Community members do not inherently find expertise in those who arrive as outsiders, whether the latter's intentions are positive, negative or unclear. Similar to how researchers subjectively reflect on participants as credible or not, researchers must establish credibility among participants throughout the process of designing, gathering, interpreting, reporting and using information. Just as interesting information becomes as powerful as data, the same information is powerful in locally relevant ways to the place and people from where it came.

In order to establish a credible exchange, it matters how researchers approach and behave with local participants in every interaction, including acknowledgment of the locals' credibility. Yung's (2001, pp2–3) description of survey development in a community illustrates the process of credibility exchange:

> *I also met with ten individuals one on one and asked what they hoped to learn from the research, what questions they would ask, who I should talk with, how the research might benefit local communities and how to get results to community members... This survey was a collaboration between myself, the researcher and a citizen's committee formed to address land use and growth in Teton County, Montana. I*

> *had been attending the meetings of this group for about one year before we began working on the survey. The purpose of the survey is to provide information that can be utilized by the citizen's committee in making recommendations about county policies.*

Community research designed and administered exclusively by experts runs counter to the community-centered approach of PAR. Community input can help to guard against experts making incorrect assumptions about the values, concerns and goals of a community whom they study, as well as allow their own bias(es) to influence research findings. Thus, for these reasons, some may argue that designing and administering research in collaboration with community members lowers the risk of irrelevance and of researcher bias influencing the outcomes, as well as increases the robustness of the findings. Without reciprocal credibility, local perceptions of research efforts may result in a belief that the outcomes will not matter and/or have missed the most critical issue(s). In more collaborative research efforts such as within a PAR model, when deciding which questions they want to ask, the biases of researchers and/or experts who fail to listen are minimized and counterbalanced, and the benefits of credibility are maximized. In this way, PAR improves the knowledge generated in a project from both community and academic perspectives. Assessment across multiple projects of the level of credibility that researchers have with community members in PAR projects, contrasted with their level of credibility in other methodologies, would facilitate a better understanding of whether PAR achieves the goal of applicable research results (Greenwood and Levin, 1998a; Levin, 2003).

Credibility also matters as a component within the very process of PAR because participants' perceptions influence what is happening and how well it is going along the way. For instance, do all of those in the process feel that their voice is heard and respected equally, or even enough? In some instances, simply being involved is not enough for community members, since 'not all participation is empowering' (Elden and Levin, 1991, p133). Hibbard and Madsen (2003) analyzed the perceptions of environmentalists involved in community-based collaborative natural resource management projects. Their analysis concludes that many environmentalists claim that their voices are not respected, valued or truly incorporated in local collaborations. From their results, many local activists did not feel the processes in which they participated were credible because of the perception that place-based collaboration can be collusions between industry and local businesspersons who benefit (Hibbard and Madsen, 2003). The activists feared that the lack of input would ultimately reduce the strength of regulations governing forest lands through decisions that favored production over the protection of these lands. Furthermore, local activist-projected new regulations would become acceptable to policy-makers and the public under the guise of local control (Hibbard and Madsen, 2003). Sturtevant and Bryan (2004) wrote a critical reply asserting that the conclusions were invalid, implying that Hibbard and Madsen conducted their research in an unethical manner. However, Hibbard and Madsen's (2004) rebuttal raised a critical point to consider when evaluating this aspect of credibility in all projects, but with a unique applicability to PAR: whose opinion of credibility matters?

Sturtevant and Bryan's (2004) focus on outcomes and on the perceptions of participants, as a whole, contrasts with Hibbard and Madsen's (2004) focus on the perceptions of a specific segment of participants and suggests that neither set of authors is entirely right or wrong. No easy answer exists for the dilemma of what level of analysis is most appropriate for considering perceptions of credibility. Instead, some attention to multiple levels of perception is required. These points relate to how Lincoln and Guba (1982, 1985) discussed issues of credibility and trustworthiness in light of attacks on naturalistic enquiry approaches. They argued that naturalistic approaches receive inordinate levels of critique about meeting the criteria of rigor assumed within positivist approaches and recommended enquiry audits as a form of establishing credibility (Lincoln and Guba, 1982; Rodwell and Byers, 1997; Creswell and Miller, 2000) within participatory designs subject to the question of whose opinion of credibility matters.

More formally, assessment of the perceptions of credibility should occur at the stakeholder group level, ideally by conducting a confidential series of interviews or administering a survey. Measuring this factor in process may also allow a negative issue to be brought to the attention of those involved in the project. Such an effort could prevent participants from becoming alienated or divided within a community amidst a PAR process. Meeting this challenge again highlights the role of researcher as facilitator, perhaps using conflict resolution tools to mediate discussion towards productive rather than divisive outcomes. Considering the effectiveness of PAR on a broader scale, it seems valuable to conduct a meta-analysis of the processes described here to determine if a particular interest group or stakeholder category consistently feels a lack of credibility across multiple PAR projects. In summary, evaluating perceptions of all participants, as Sturtevant and Bryan (2004) suggest, is an important component in any PAR assessment.

Continuity of trust

In this context, credibility between participants and researchers also enhances the quality and integrity of the methodological approach(es) that a community and research group may decide to use. As facilitators, PAR researchers provide structure, guidance and technical input (if needed and where necessary) to community-centered research efforts. Sustaining the relationship between facilitation and community ownership requires a continuity of trust between those involved. The continuity of trust is the joint responsibility of those entering into a research activity that may very well have undefined roles and expectations.

Relationships of trust matter a great deal in scenarios of negotiation, as found with community-centered research (Udas, 1998). A lack of trust may inhibit progress if a common direction is not identified. Trust forms the mortar of a foundation between all project participants and enables a project to endure. In this way, the level and continuity of trust within a project will likely affect the levels and type of participation in which community members will engage.

Related to trust between participants and researchers, trustworthiness of the data also matters a great deal to the PAR process given comparison and contrasts to more conventional scientific paradigms. Pretty (1995) noted frequent critiques of PAR methods and findings as sloppy, lacking rigor or biased. Authenticity and trustworthiness criteria (i.e. if participation changes individuals, and increases awareness and action prompted by the process; see Lincoln, 1990; Pretty, 1994, 1995) prove instructive and pertinent to PAR evaluation and the domain of trust. For PAR, the points of validity and reliability rely on attention to the processes of the research to ensure that the data and results are worthy and meaningful to project participants.

Criterion two: Reciprocal production of knowledge

Reciprocity in the joint pursuit of knowledge is the second criterion essential to evaluate PAR. By reciprocity, we refer to the need for participatory research to fulfill community goals, as well as those of the research-oriented individual or organization. Within a PAR model, if research is truly community-centered, it makes community needs primary and the organizational/institutional needs secondary. However, the process of academic enquiry within a PAR model becomes one and the same with the effort to meet community needs. These goals need not be perceived as distinct, competitive or counter to one another, as they are in traditional positivistic models of research (Whyte, 1989; Hughes, 2003). Instead, the goals complement one another because they allow for an ongoing negotiation to balance multiple needs. In turn, such a process may minimize the effects of social marginalization, or the feeling and perception that one's social status is compromised in relation to others, within PAR processes and outcomes.

Overcoming marginalization

In the context of a reciprocal approach, an important aspect of PAR is to ensure that marginalized groups in communities have a voice in the research process (Nussbaum et al, 2004). By voice, we mean something more substantive than the formal input characteristics of many public hearing formats. In our use here, marginalized refers broadly either to any group who has either seen little opportunity to access resources and empower themselves to identify solutions to a problem, or were perhaps subject to an approach that provided a traditional solution with little effective implementation. Greenwood and Levin (1998a) also employ the term marginalized more broadly than referring only to disenfranchised or oppressed groups, and suggest that PAR itself is marginalized in certain contexts because it interrupts the social structures of inaction intrinsic to more traditional approaches.

Inclusive processes give people creative time and space for interaction. That interaction can become an opportunity to generate knowledge and examine their own situations, and empower citizen and community groups to use knowledge to engage in action that will improve their quality of life (Finn, 1994; Udas, 1998). Including marginalized groups in PAR processes opens the possibility of mutual respect and a sense of credibility between participants and researchers. This reciprocity also supports the need for researchers to develop a sense of confidence in community participants, in turn affecting judgment and the interpretation of data for analyses. We emphasize this point to stress the relationships essential to successful PAR projects that produce trustworthy information for communities and academics alike (Lindsey and McGuinness, 1998). Communities benefit from such a structure because collaboration on PAR projects can bridge divisions in communities (Brown, 1985). Breaking down barriers between diverse segments of a community enhances a sense of well-being and connectedness among participants. This, in turn, can lead to both more inclusive research and broader acceptance of its findings, as well as the expanded and more applicable outcomes that result from their application.

But politically, on what basis is a participant's status determined and who decides – the participant, other participants or the researcher/s? This important question for PAR facilitation goes beyond simply ensuring broad involvement. In collaborative processes, sensitivity to how social power may affect participants' input and how ideas and comments are received matters for successful facilitation. Recognizing how labels such as 'marginalized' and the power relationships among community members affect PAR processes is critical in this regard.

For example, in the case of community forestry issues in the US, the inclusion of marginalized groups may develop a reciprocal voice that had no prior means to be heard. Ballard (2001) conducted a participatory research project for the sustainable management of non-timber forest products. The project emphasized development of a participatory monitoring program that would contribute to bettering livelihoods for a number of harvesters representing a variety of different cultural and ethnic backgrounds in Washington State. Ballard's (2001, p9) case highlights the effect that PAR can have when a local group with little voice in the community or little impact upon policy decisions, such as natural resource management, becomes a participant with the opportunity of having their work and perspectives legitimized by a less marginalized entity:

> *An indirect benefit of working with me for the harvesters is that nearly every step in the research project so far has also involved Forest Service managers and/or researchers, thereby providing an informal exhibition*

> *of harvester knowledge, experience and concern for sustainability to the Forest Service District and forest supervisors, as well as the state DNR [Department of Natural Resources] personnel, with recommendations for permitting and management changes based on our work. This will hopefully also expose managers and decision-makers to the range of harvesters' knowledge and experience that can contribute to future forest management practices and policies affecting non-timber forest products and harvester livelihoods.*

However, in some situations, marginalization is more difficult to discern. For example, do western ranch families, many of whom remain extremely independent and on the verge of displacement, constitute a marginalized group? Gripne (2005) investigated the utility of grassbanks as a conservation tool to address this complex issue. Social and economic dimensions of the project highlighted how characteristics of a constituency's ties to a given land-use policy may constitute equity and fairness in decision-making equations to determine the extent of marginalization. Today, many ranchers might readily testify that they feel marginalized in a world where environmentalists have developed increasingly strong legal voices, tools and positions of intolerance about the impacts of livestock grazing and other traditional uses in working landscapes. In order to stay in business and remain economically viable, some ranchers feel the pressure to subdivide their ranches, often accompanied by a feeling of further disenfranchisement about community well-being and decision-making. In this setting, having a voice in a research process becomes a means of investigating the options for changing not only one's livelihood, but also the social identities and attachments long a part of having control over one's own destiny.

Some have intimated (Donahue, 1999; Hewitt, 2002) that, historically, ranchers have exercised a disproportionate amount of power through social ties and political influence on decision-makers and land managers. From such a perspective, ranchers may now experience a shift to a more equitable share of that power. But a variety of accounts on the complexities of livestock grazing, rangeland health and the politics of resource management affect which (and to what degree) groups feel marginalized. This point is critical to understanding the relationship between the trust aspect of community control and reciprocal scholarship. Without a basis of trust, independently minded individuals and groups, such as ranchers, may not embrace a research project that they sense may further the marginality of the position in which they already perceive themselves.

While identification of marginalized groups in a community can be challenging, evaluating this dimension of PAR projects requires considering whose participation is missing. Once engaged, people must also provide their assessment of credibility for the process and other participants. The only means to address this difficult issue is for a research team to design an open process that not only fosters inclusiveness, but also allows for flexibility when the need arises according to local perspectives.

Only when community members develop involvement and assume an active role that includes giving feedback about the process in which they are engaged can trust about the research process be generated.

Issues of marginalization also raise the question of fair representation. Essentially, who holds the rights, in a community-based research project, to represent the views of a collective? Individuals can serve as representatives, but must rely on the reciprocity of trust and perceived balance of expectations and roles of responsibility to have their voice/s remain credible. As such, PAR may often be more easily facilitated through established community organizations that already have a continuity of trust and a basis for relations with others. These structures, while beneficial, may also lead to other types of problems about who is and who is not involved related to perceptions of who holds the power and knowledge (Gaventa and Cornwall, 2001). The key task as PAR researchers dealing with issues of marginalization is to respect various definitions of marginalization, to attempt to include a diverse set of participants in the research process, and to maintain and enhance aspects of credibility once they are included.

Knowledge production

In terms of the need for social relationships to make PAR successful, Wilkinson (1991) emphasized that community is an interactive process involving attachment to place. In this context, the co-production of knowledge contributes to the ongoing process of creating community and produces better theory by grounding that knowledge in local realities.

Contextual knowledge often exists within participants' sense of place, such as the ideas about a community setting or natural resource that participants in the PAR process may share. All stakeholders, however, do not necessarily share a common sense or definition of community. For example, if the management of public forest resources is an issue in the community, non-resident stakeholders also may get involved. Even if all involved have a sense of place about the forest and its uses, however, a facilitated process such as PAR may highlight fundamentally different perspectives for management. In this context, trustworthy knowledge, as well as a process that participants can rely on as credible, is fundamental to cooperative agreement. It is possible that despite differences in opinion, many stakeholders do share connections to particular places in the landscape. Building on these commonalities can begin to facilitate the appreciation of local knowledge or build mutual respect for different knowledge systems. In turn, these benefits can lead to the development of research processes that produce information with a high degree of face validity for PAR participants and academics alike by virtue of their inclusiveness and broad acceptance across stakeholder groups.

As noted for PAR, participants share control of the research process in different aspects of the project. Knowledge production starts with project initiation and the identification of research questions. Those seeds or motivations must originate somewhere through a catalyst (Maguire, 1993), whether internal or external to the community. Commitment to a PAR approach requires an understanding about asserting ideas too strongly or with poor timing in light of the first criterion: community control. As such, researchers may gain entry to possible projects with serendipitous and unplanned opportunities that arise. Thus, in addition to sharing power, the idea of knowledge production as a reciprocal effort means that synergistic effects from working together will produce accessible and more comprehensive outcomes.

To assess the perceptions of those involved in the PAR process, the methodology should identify participants' attitudes about the degree of control over the research they perceive having. In addition, a determination of what specific research tasks community members were heavily involved in and a consideration of how many community members were involved and what groups were represented is also appropriate. Determining the breadth of community members' involvement across tasks in the research process is particularly important because it may be tempting for researchers to rely on participants to help design and conduct research, but to delve into analysis themselves. Nevertheless, in many cases, only when participants also work to analyze the information collected will it be trusted by community participants. More importantly, however, only when that trust exists will the findings be applied to achieve outcomes. Researchers must also keep their purposes in mind to analyze the effort as a whole and, in the context of other cases and literature, to evaluate them on the broader scale.

Reason (1994) noted that many participatory research projects depend upon someone with the initiative, time, skill and commitment. However, community activists point out that overcoming the struggles of reciprocal scholarship remains challenging because of perceptions of scientists as self-serving and unwilling to take political risks or to engage in vigorous civic dialogues that community participants feel are needed (Ward, 2000). Inversely, citizens often have explicit political agendas that aim to change public policy and to critique scientists on their blindness towards biases that shape priorities and purpose. These points provide the context for a discussion of our third criterion, which focuses on PAR products and outcomes.

CRITERION THREE: OUTCOMES AND WHO BENEFITS

Translating the knowledge production into action constitutes a third key criterion for evaluations of participatory action. In other words, generating and discovering knowledge in communities makes little practical sense if that information is not also put to use for social change in the form of practical outcomes. PAR literature has demonstrated repeated examples of this essential element (O'Looney, 1998; Schafft and Greenwood, 2003; Nussbaum et al, 2004); however, from a practical standpoint the development of action is often case specific with regard to content and context. Thus, PAR may occur more as a custom model adapted to situations and groups for their utility and benefit, rather than a set formula with known or predictable outcomes. Given this, evaluation of PAR projects entails continual contact with a community, involving identification of how the information collected has been used, as well as an assessment of whether the efforts meet the community goals and objectives identified during the early stages of the research process.

Utilizing knowledge to build community capacity

A related concept for consideration when evaluating outcomes is whether the research process has enhanced the community's capacity to engage in research and work with new or additional information. As noted, according to some PAR scholars, one of the most important benefits of PAR is the empowerment that community members involved in the research obtain from the experience (Wismer, 1999; Mordock and Kransy, 2001). When evaluating PAR research, assessing these capacities can help to determine whether these effects have taken place and can ascertain the breadth of outcomes from the process.

Should all knowledge production and utilization involve participants? In other words, is this a participatory research ideal or requirement? Is the researcher allowed to publish separate conclusions? What happens when participant and researcher conclusions remain in conflict? It is possible that such a situation, even if unpleasant, is not detrimental to strong scientific research as the diverse uses for the information collected may allow those involved to agree to disagree (Belsky, 2004). If community control is truly exercised, it is not the researcher's place to determine the truth of research, but, instead, to facilitate its application in dealing with community issues. Similarly, it is within a researcher's purview to determine how results apply in efforts to generate academic knowledge, and such a focus may be a contributing factor to the development of the different conclusions drawn. In short, looking at the reasons behind different conclusions may be helpful in understanding them; but the differences in and of themselves do not necessitate the conclusion of a failed or unproductive project.

Institutional program constraints and challenges in using PAR

The result of combining the open-ended PAR approach with strict academic requirements often results in research that may have elements of a PR approach, yet at the same time may remain inconsistent with other aspects of the method. Determining how the academic community evaluates findings from PAR is also an important factor to consider in the differentiation between PAR and PR outcomes. Such differences may be particularly important in instances where community control strongly took shape. In such a situation, many researchers may have approached a community research design with participation-oriented ideals. Achieving project objectives, however, with the same original ideas and approaches remains an emergent process rather than a prescription. In PAR, the ultimate measured outcome is if the community participants feel that they benefited from the project. However, for academic performance, emphasis on community benefits as outcomes will often not produce the types of research products preferred for evaluation (i.e. refereed journal articles using deductive approaches).

The traditional approach to gaining funding for graduate research also presents multiple problems for academic advisers and students attempting participatory research. The first potential obstacle occurs in the pursuit of funding. While positivist approaches – those that emphasize the validity of knowledge based on the scientific method – have come under increasing attack in the past few decades, they constitute the dominant research paradigm receiving institutional support (Schroeder, 1997). The CFERP program is one of the few funding mechanisms that specifically promotes participatory research in the context of natural resources and does not require *a priori* hypotheses and a detailed study design prior to allocating funds. Most funding sources require a study design with clearly articulated hypotheses that are the antithesis of the open-approach participatory approaches. PR does not lend itself to developing *a priori* hypotheses for dissertation proposals or grants to fund research.

In addition to the problem of a rigid, positivistic proposal format that most funding sources adhere to, time-limited grants and two- to four-year windows allotted to most MSc and PhD programs are rarely conducive to the time scale needed for participatory research projects. Inexperienced researchers and, especially, students feeling the pressure of funding often underestimate the time investment needed to conduct a high-quality study using a participatory approach (Berardi, 2002). Successful entry into the community can take a significant amount of time and is just the first step in a long process of engaging in participatory research. Thus, the open-ended needs of participatory approaches may become barriers within the requirements of many traditional funding sources (Maguire, 1993; Park, 1993; Yung, 2001).

Conclusion: The design of PAR and measures of its utility

Despite our outline of criteria, the field of PR faces a dilemma: as an open-ended approach, PR may run counter-intuitive to developing standardized criteria or a metric to rate the quality of research efforts or outcomes at the various levels identified and discussed. As noted earlier, enquiry audits offer a means of systematically reviewing naturalistic and participatory designs to ensure that results are trustworthy and dependable. However, in the absence of such a step, important lessons in the learning and doing of research rely heavily on action, compassion and the exchange of knowledge.

As such, our goal here was not to establish criteria by which to judge research as a success or failure. Nor was it intended to establish a benchmark of acceptability, standards or normative expectations. The challenge here, rather, was to look critically at whether PAR defies evaluation altogether, or if integral components exist in similar ways in some community projects in order to identify patterns. This work concludes that such key elements do exist and that these similarities can be considered for use as criteria in evaluating PAR projects and for application in the larger methodological approach. Thinking about issues of evaluation in the planning stages of a project is a useful way for researchers to orient themselves towards important tasks and towards the key issues to be considered and addressed. This chapter outlines several key categories for consideration as we seek to define, implement and evaluate any project.

In summary, assessment of PAR includes some evaluation of each of the following: whether a fundamental understanding of PAR approaches, the intent to engage in those ideals, or the accomplishment of a participatory process matters the most. We assert that some principles identified here – community-centered control, reciprocal knowledge production and an orientation focused on building community capacity in outcomes – will facilitate PAR, while others will not. Researchers who are new to the participatory research approach may benefit from having these fundamental ideals outlined. However, our attempt to do so should not be taken as comprehensive or as having universal applicability. Our articulation here emerges in response to recent literature, discussion and activity of heightened interest in PAR and more general participatory approaches. Our underlying goal is to further stimulate the development of measures, understanding and evaluation tools for PAR. Our hope is that as work built on PAR principles continues to develop, the ideas discussed here can provide further guidance to the creativity of these endeavors.

REFERENCES

Bailey, D. (1992) 'Using participatory research in community consortia development and evaluation: Lessons from the beginning of a story', *The American Sociologist*, vol 23, no 4, pp71–82

Baker, M. and Kusel, J. (2003) *Community Forestry in the United States: Learning from the Past, Crafting the Future,* Island Press, Washington, DC

Ballard, H. (2001) *Community Forestry and Environmental Research Partnerships Program Final Report,* www.cnr.berkeley.edu/community_forestry/People/Final%20Reports/ballard_report.pdf, accessed 28 July 2003

Beck, U. (1995) *Ecological Enlightenment: Essays on the Politics of Risk Society*, Humanities Press, Englewood, NJ

Belsky, J. M. (2004) 'Reflections on ecotourism research in Belize: Implications for critical qualitative methodology in tourism research', in J. Phillmore and C. Goodson (eds) *Qualitative Methods in Tourism Research*, Routledge Press, Oxford, pp274–291

Berardi, G. (2002) 'Commentary on the challenge to change: Participatory research and professional realities', *Society and Natural Resources*, vol 15, pp847–852

Brown, D. L. (1985) 'People centered development and participatory research', *Harvard Educational Review*, vol 55, no 1, pp69–75

Caldwell, J. Y., Davis, J. D., Du Bois, B. Echo-Hawk, H., Erickson, J. S., Goins, R. T., Hill, C., Hillabrant, W., Johnson, S. R., Kendall, E., Keemer, K., Manson, S. M., Marshall, C. A., Running Wolf, P., Santiago, R. L., Schacht, R. and Stone, J. B. (2005) 'Culturally competent research with American Indians and Alaska natives: Findings and recommendations of the First Symposium of the Work Group on American Indian Research and Program Evaluation Methodology', *The Journal of the National Center*, vol 12, no 1, pp1–21

Carr, A. and Wilkinson, R. (2005) 'Beyond participation: Boundary organizations as a new space for farmers and scientists to interact', *Society and Natural Resources*, vol 18, pp255–265

Conchelos, G. and Kassam, Y. (1981) 'A brief review of critical opinions and responses on issues facing participatory research', *Convergence*, vol 14, no 3, pp52–64

Cornwall, A. and Jewkes, R. (1995) 'What is participatory research?', *Social Science Medicine*, vol 41, no 12, pp1667–1676

Crespo, I., Palli, C. and Lalueza, J. L. (2002) 'Moving communities: A process of negotiation with a Gypsy minority for empowerment', *Community, Work and Family*, vol 5, no 1, pp49–66

Creswell, J. W. and Miller, D. L. (2000) 'Determining validity in qualitative inquiry', *Theory into Practice*, vol 39, no 3, pp124–130

Crowfoot, J. E. and Wondolleck, J. M. (1990) *Environmental Disputes: Community Involvement in Conflict Resolution*, Island Press, Washington, DC

David, M. (2002) 'Problems of participation: The limits of action research', *International Journal of Social Research Methodology*, vol 5, no 1, pp11–17

Donahue, D. L. (1999) *The Western Range Revisited: Removing Livestock from Public Lands to Conserve Native Biodiversity*, University of Oklahoma Press, Norman, OK

Eisenhauer, B. and Wulfhorst, J. D. (2005) 'Capitalizing on the potential to empower and mobilize', *Community-centered Research Series: Measuring What Matters*, CCR6, Western Rural Development Center, Logan, UT, http://extension.usu.edu/files/publications/publication/pub_5991888.pdf, accessed 12 December 2006

Elden, M. and Levin, M. (1991) 'Cogenerative learning: Bringing participation into action research', in W. F. Whyte (ed) *Participatory Action Research*, Sage, Newbury Park, pp127–142

Finn, J. (1994) 'The promise of participatory research', *Journal of Progressive Human Services*, vol 5, no 2, pp25–42

Fischer, F. (2000) *Citizens, Experts and the Environment: The Politics of Local Knowledge*, Duke University Press, Durham, NC

Flora, C. B., Larrea, F., Ehrhart, C., Ordonez, M., Baez, S., Guerrero, F., Chancay, S. and Flora, J. L. (1997) 'Negotiating participatory action research in an Ecuadorian sustainable agriculture and natural resource management program', *Practicing Anthropology*, vol 19, no 3, pp20–25

Foucault, M. (1980) *Power/Knowledge: Selected Interviews and Other Writings*, Vintage Books, New York, NY

Gaventa, J. and Cornwell, A. (2001) 'Power and knowledge', in P. Reason and H. Bradbury (eds) *Handbook of Action Research Participative Inquiry and Practice*, Sage, London, pp70–80

Greenwood, D. J. and Levin, M. (1998a) 'Action research, science and the co-optation of social research', *Studies in Cultures, Organizations and Societies*, vol 4, pp237–261

Greenwood, D. J. and Levin, M. (1998b) *Introduction to Action Research: Social Research for Social Change*, Sage, Thousand Oaks, CA

Greenwood, D. J., Whyte, W. F. and Harkavy, I. (1993) 'Participatory action research as a Process and as a Goal', *Human Relations*, vol 46, no 2, pp175–192

Gripne, S. (2005) 'Grassbanking: Bartering for conservation', *Rangelands* vol 27, no 1, pp24–26

Guevara, J. R. Q. (1996) 'Learning through participatory action research for community ecotourism planning', *Convergence*, vol 29, no 3, pp24–40

Hewitt, W. L. (2002) 'The "cowboyification" of Wyoming agriculture', *Agricultural History*, vol 76, no 2, pp481–494

Hibbard, M. and Madsen, J. (2003) 'Environmental resistance to place-based collaboration in the US West', *Society and Natural Resources*, vol 16, pp703–718

Hibbard, M. and Madsen, J. (2004) 'Response to Sturtevant and Bryan', *Society and Natural Resources*, vol 17, pp461–466

Hughes, J. N. (2003) 'Commentary: Participatory action research leads to sustainable school and community improvement', *School Psychology Review*, vol 32, no 1, pp38–43

Levin, M. (2003) 'Action research and the research community', *Concepts and Transformations*, vol 8, no 3, pp275–280

Lincoln, Y. S. (1990) 'The making of a constructivist: A remembrance of transformations past', in E. G. Guba (ed) *The Paradigm Dialog*, Sage, Newbury Park, CA, pp67–87

Lincoln, Y. S. and Guba, E. G. (1982) 'Establishing dependability and confirmability in naturalistic inquiry through an audit', Paper presented at the 66th Annual Meeting of the American Educational Research Association, 19–23 March, New York, NY

Lincoln, Y. S. and Guba, E. G. (1985) *Naturalistic Inquiry*, Sage, Newbury Park, CA

Lindsey, E. and McGuinness, L. (1998) 'Significant elements of community involvement in participatory action research: Evidence from a community project', *Journal of Advanced Nursing*, vol 28, no 5, pp1106–1114

Maguire, P. (1993) 'Challenges, contradictions and celebrations: Attempting participatory research', in P. Park, M. Brydon-Miller, B. Hall and T. Jackson (eds) *Voices of Change: Participatory Research in the US and Canada*, Bergin and Garvey, Westport, CT, pp157–176

McTaggart, R. (1991) 'Principles for participatory action research', *Adult Education Quarterly*, vol 41, no 3, pp168–187

Medley, K. E. and Kalibo, H. W. (2007) 'Global localism: Re-centering the research agenda for biodiversity conservation', *Natural Resources Forum*, vol 31, pp151–161

Michaels, S., Mason, R. J., Solecki, W. D. (2001) 'Participatory research on collaborative environmental management: Results from the Adirondack Park', *Society and Natural Resources*, vol 14, pp251–255

Moote, M., Brown, B., Kingsley, E., Lee, S. X., Marshall, S., Voth, D. and Walker, G. B. (2001) 'Process: Redefining relationships', in G. J. Gray, M. J. Enzer and J. Kusel (eds) *Understanding Community-based Forest Ecosystem Management*, The Haworth Press, Inc, New York, NY, pp97–116

Mordock, K. and Kransy, M. E. (2001) 'Participatory action research: A theoretical and practical framework for EE', *The Journal of Environmental Education*, vol 32, no 3, pp15–20

Norgrove, L. and Hulme, D. (2006) 'Confronting conservation at Mount Elgon, Uganda', *Development and Change*, vol 37, no 5, pp1093–1116

Nussbaum, R. H., Hoover, P. P., Grossman, C. M. and Nussbaum, F. D. (2004) 'Community-based participatory health survey of Hanford, WA Downwinders: A model for citizen empowerment', *Society and Natural Resources*, vol 17, pp547–559

O'Looney, J. (1998) 'Mapping communities: Place-based stories and participatory planning', *Journal of the Community Development Society*, vol 29, no 2, pp201–236

Pain, R. and Francis, P. (2003) 'Reflections on participatory research', *Area*, vol 35, no 1, pp46–54

Park, P. (1993) 'What is participatory research? A theoretical and methodological perspective', in B. E. Hall, P. Park, M. Brydon-Miller and T. Jackson (eds) *Voices of Change: Participatory Research in the US and Canada*, OISE, Toronto, pp1–20

Park, P. (1999) 'People, knowledge and change in participatory research', *Management Learning*, vol 30, no 2, pp141–157

Phadke, R. (2005) 'People's science in action: The politics of protest and knowledge brokering in India', *Society and Natural Resources*, vol 18, pp363–375

Powers, J. Cumbie, S. A. and Weinert, C. (2006) 'Lessons learned through the creative and iterative process of community-based participatory research', *International Journal of Qualitative Methods*, vol 5, no 2, pp1–9, www.ualberta.ca/~ijqm/backissues/5_2/pdf/powers.pdf, accessed 15 June 2007

Pretty, J. N. (1994) 'Alternative systems of inquiry for sustainable agriculture', *IDS Bulletin*, vol 25, no 2, pp37–48

Pretty, J. N. (1995) 'Participatory learning for sustainable agriculture', *World Development*, vol 23, no 8, pp1247–1263

Purnomo, H., Mendoza, G. A. and Prabhu, R. (2004) 'Model for collaborative planning of community-managed resources based on qualitative soft systems approach', *Journal of Tropical Forest Science*, vol 16, no 1, pp106–131

Quaghebeur, K., Masschelein, J. and Nguyen, H. H. (2004) 'Paradox of participation: Giving or taking part?', *Journal of Community and Applied Social Psychology*, vol 14, pp154–165

Reason, P. (1994) 'Three approaches to participative inquiry', in N. K. Denzin and Y. S. Lincoln (eds) *Handbook of Qualitative Research*, Sage, Thousand Oaks, CA, pp324–329

Reason, P. and Bradbury, H. (eds) (2001) *Handbook of Action Research: Participative Inquiry and Practice*, Sage, London

Rodwell, M. K. and Byers, K. V. (1997) 'Auditing constructivist inquiry: Perspectives of two stakeholders', *Qualitative Inquiry*, vol 3, no 1, pp116–134

Schafft, K. A. and Greenwood, D. J. (2003) 'Promises and dilemmas of participation: Action research, search conference methodology and community development', *Journal of the Community Development Society*, vol 34, no 1, pp18–35

Schroeder, K. (1997) 'Participatory action research in a traditional academic setting: Lessons from the Canada Asia Partnership', *Convergence*, vol 30, no 4, pp41–49

Sillitoe, P. (2006) 'Ethnobiology and applied anthropology: Rapprochement of the academic with the practical', *Journal of the Royal Anthropology Institute*, pp119–142

Simonson, L. J. and Burshaw, V. A. (1993) 'Participatory action research: Easier said than done', *The American Sociologist*, vol 24, no 1, pp27–38

Sims, B. and Bentley, J. (2002) 'Participatory research: A set of tools, but not the key to the universe', *Culture and Agriculture*, vol 24, no 1, pp34–41

Sturtevant, V. and Bryan, T. (2004) 'Commentary on "Environmental resistance to place-based collaboration" by M. Hibbard and J. Madsen', *Society and Natural Resources*, vol 17, pp455–460

Udas, K. (1998) 'Participatory action research as critical pedagogy', *Systemic Practice and Action Research*, vol 11, no 6, pp599–628

Vio Grossi, F. (1981) 'Socio-political implications of participatory research', *Convergence*, vol 14, no 3, pp43–51

Ward, C. (2001) *Canaries on the Rim: Living Downwind in the West*, Verso Press, New York, NY

Weber, E. P. (2000) 'A new vanguard for the environment: Grass-roots ecosystem management as a new environmental movement', *Society and Natural Resources*, vol 13, pp237–259

Whyte, W. F. (1989) 'Advancing knowledge through participatory action research', *Sociological Forum*, vol 4, no 3, pp367–385

Whyte, W. F. (ed) (1991) *Participatory Action Research*, Sage, Newbury Park, CA

Whyte, W. F. (1998) 'Rethinking sociology: Applied and basic research', *The American Sociologist*, vol 29, no 1, pp16–19

Wilkinson, K. (1991) *The Community in Rural America*, Greenwood Press, Westport, CT

Wismer, S. (1999) 'From the ground up: Quality of life indicators and sustainable community development', *Feminist Economics*, vol 5, no 2, pp109–114

Yung, L. (2001) 'Some Lessons from participatory research', *Regeneration: Newsletter of the Community Forestry and Environmental Research Partnerships Program*, vol 1, no 1, pp5–8, www.cnr.berkeley.edu/community_forestry/Community/newsletter/Newsletter_summer_2001.pdf, accessed 14 July 2003

3

Challenges to Institutionalizing Participatory Research in Community Forestry in the US

Carl Wilmsen and Ajit Krishnaswamy

INTRODUCTION

I later began to realize that by engaging in conversations with disputants, I was asking them *what was important and meaningful about the conflict. From there I began to formulate research questions... Finally, I began to sense that through continued conversations with participants, I was indirectly engaging them in a participatory process of data gathering, analysis and interpretation, which fit quite naturally with my interest in exploring social understanding and meaning. Only in retrospect does it appear that I followed, albeit indirectly, a participatory research approach.* (Bryan, 2002)

The essential incentive for me [to be involved] is the research itself. The Steering Committee members have changed the shape of Shannon's research ... I feel that we have helped to set a higher standard for restoration as the cultural uses of the natural resources require an extremely healthy, managed landscape. Shannon has a more interesting set of questions and possibilities than she did at the outset of her work. Her answers will mean something to the mining community, native tribal peoples, restoration ecologists, scholars and those visiting schoolchildren who may be our future policy-makers ... I have lived in the Cache Creek watershed for most of my life. At last, I have an invitation to be part of the solution to problems of which I have been peripherally aware for over 30 years. It feels good to be involved with something so central to the healing of the land and the people. (Ross, 2003)

These two statements describe different degrees of community engagement in research. The first suggests that the researcher deemed his research participatory because he developed his conclusions through an iterative process of hypothesis generation, information gathering, reflection and hypothesis revision. The second statement indicates full engagement of community members in the research and a resulting increase in their ability to participate in resource management in ways that they find more fulfilling.

Together, the two statements represent the promise and the challenge of institutionalizing participatory research in community-based natural resource management (CBNRM) in the US. On the one hand, institutionalizing participatory research will make it commonplace; its principles will be incorporated within standard research practice. This promises to reorient research practice so that it promotes mutual learning among community members and professional researchers, advances scientific knowledge and increases community well-being. On the other hand, institutionalization may involve entrenching, in practice, contradictory pressures that sustain uncertainty about how to balance community needs with research imperatives. In the worst case scenario, it may even involve perversion of participatory research so that its principles may be incorporated within the practices of government agencies, economic development organizations and universities in ways that further the objectives of those agencies and organizations more so than to meet community needs.

Contradictory pressures in the practice of participatory research may lead to greater emphasis being placed on the goals of the professional researcher(s) even when intentions are good. These pressures stem from a fundamental separation of research and community development in the current educational system and include career incentives, structural factors in education, funding priorities and funding agency requirements that pull researchers in different directions. In this chapter we examine how these pressures affected two planned efforts to further participatory research's institutionalization in US community forestry. The Community Forestry and Environmental Research Partnerships (CFERP)[1] program was established in 1996 at the University of California, Berkeley, and the National Community Forestry Center (NCFC) operated from 2000 until 2004 as a project of the National Network of Forest Practitioners. Both were established to promote participatory research and community forestry: CFERP in the academy and the NCFC in non-profit organizations and communities.

We define institutionalization of participatory research as the process of integrating it within the regular procedures of an organization so that it becomes a standard feature of research practice. Participatory research is being institutionalized to a certain extent in several fields in the US. It has been practiced in education, public health and community design in the US for several decades (Park, 1993;

Sanoff, 2000) and it is currently being institutionalized within CBNRM as well. Although some organizations devoted to applying participatory research to environmental issues, such as the Citizens Clearinghouse for Hazardous Waste, have been around since the 1970s, a number of new organizations addressing environmental issues with participatory research have recently been established. In addition, individual scholars who teach and practice participatory research are scattered among universities and colleges across the country. In 1998, the Loka Institute identified 40 university-based centers for participatory research in the US (Sclove et al, 1998), and today, other efforts, such as the National Community-based Networking Initiative, are continuing to expand the number of campuses, faculties and students engaging in or supporting participatory research.

Contradictory pressures constitute a major issue in the institutionalization of participatory research. They are distinct from co-option, another major issue, which is practice that enlists community members in projects and activities of sponsoring organizations that are intended to advance the agenda of the sponsoring agency more than to advance community well-being. For the past several years international development organizations such as the World Bank have required the participation of community members in the community development work that they sponsor. The World Bank has been criticized, however, for using participatory methodologies and practitioners to enforce its own neo-liberal development agenda (Cooke, 2004). This is an agenda that seeks free-market solutions to the problems of poor people without attending to the power imbalances that effectively exclude the poor from meaningful involvement in addressing the issues that directly affect their lives and livelihoods. The co-option of community members through participatory activities in this context is deliberate.

Contradictory pressures, on the other hand, result from current structures, expectations and assumptions in the educational system. They do not necessarily entail deliberate attempts to co-opt people into particular agendas, although this remains a risk. The NCFC and CFERP have both encountered significant challenges stemming from these contradictory pressures in their efforts to institutionalize participatory research. After saying a few words to introduce the NCFC and CFERP, we will discuss these issues in depth.

THE NATIONAL COMMUNITY FORESTRY CENTER (NCFC)

The NCFC was one of the few institutionalized efforts in the US geared specifically towards developing the capabilities of communities to engage in their own land stewardship through participatory research. It was established on the premise that the capacity of local communities to participate in forest management and policy is limited by their lack of access to information and that incorporating local knowledge in research is the key to better forest management.

The NCFC was established in 2000 by a four-year grant from the US Department of Agriculture's (USDA) Cooperative State Research Extension and Education Service (CSREES) to the National Network of Forest Practitioners (NNFP), a non-profit national forum of groups and individuals involved in community forestry. Until its closure in 2004, the NCFC was a decentralized network of four regional centers located in the Southwest, Southeast, Northeast and Pacific West regions. The NCFC regional centers adopted three strategies to support forest communities. First, using participatory research, each center worked collaboratively and intensively in partnership with two to five communities every year to build their capacity to conduct research. Second, regional center researchers conducted biophysical and social science research on topics relevant to communities in the region. Third, regional centers provided technical assistance and training to communities.

The Community Forestry and Environmental Research Partnerships (CFERP)

The CFERP program was created in 1996 to fulfill a critical need for research on community forestry institutions and the processes that support them in the US. At that time, while much research had been done on community forestry efforts in other countries, little had been conducted in the US on the organizing efforts of forest communities to gain some control over the management of forest lands adjacent to them. The mission of the program is to nurture a new generation of scholars and university–community partnerships to build scholarly and community capacity for stewardship of natural resources in ways that are socially just, environmentally sound and economically sustainable. The program supports graduate student-led participatory research as the primary means of achieving its mission. Research is intended to enhance understanding of the conditions under which community-based natural resource management thrives in the US, as well as to build capacity in communities for stewarding natural resources and for greater self-determination.

Promoting institutionalization of participatory research in the CFERP and NCFC

CFERP and the NCFC each built mechanisms into their programs and organizational structures to facilitate the shared decision-making that participatory research entails. CFERP has concentrated on educating graduate students, academic faculty, community members, government officials and other professionals about the benefits of participatory research and what a participatory approach entails through granting fellowships, as well as through its annual workshop, publications, fellow presentations at conferences and networking.

One innovation is the granting of pre-dissertation fellowships. Pre-dissertation fellowships provide graduate students with a small stipend to enable them to spend time with community members while still designing their research to learn what the community's research needs and desires are. The intent is to have the students develop their research questions in conjunction with community members. Doing so places the student in the position of a research facilitator and empowers community members to shape the research agenda. This is a key aspect of participatory research since research questions and agendas developed by the community are more likely to produce information relevant to the local situation than ones developed entirely by professional researchers.

The NCFC promoted community involvement and shared decision-making through the creation of Bioregional Advisory Councils (BACs). The BACs represented diverse citizen interests, and guided the work of each regional center by helping them to select partner communities and set research priorities. Each BAC was charged with making sure that regional center research was actually participatory and involved the local community. The centers were committed to developing effective, strong and representative BACs. While constituting the BACs, the centers tried to maintain a balance among advocates, policy people, experts and researchers, on the one hand, and practitioners, on the other. At least half of the BAC membership was expected to represent rural communities of place or forest workers.

INSTITUTIONAL BIASES AGAINST ADOPTING PARTICIPATORY APPROACHES

The mechanisms that CFERP and the NCFC put into place for facilitating joint decision-making were sound; but their effectiveness has been hampered by countervailing forces. The separation of research and community development affected the efforts of both organizations in the way in which their activities meshed with existing structures, participant and funder expectations, and career incentives within and between the university and the non-profit sectors.

In contrast to the separation of research and community development, participatory research rests on the notion that the two are complementary. In participatory research, theory and practical applications are seen as integral to one another. Practical action produces the knowledge of the everyday world that informs theory, and theory serves as a guide to improving practical action. This approach contrasts with the traditional approach to research in which practical action is avoided as part of the effort to maintain objectivity. Proponents of participatory research face many challenges in attempting to change that traditional approach. The academy is a central component of the accepted system through which the legitimacy of knowledge is determined; as such, it promotes and defends specific structures and practices that have definite implications not only for the way in which research is conducted, but also for the way in which new researchers are trained.

Significant barriers to accepting participatory research in the academy include the way in which graduate education is structured and the incentives for advancement in one's career. Under current rules governing the composition of thesis committees, graduate education is not structured to provide students with opportunities to develop their research questions in conjunction with community members. Students are expected to do research that engages current theoretical debates in their chosen field of study, often ignoring or de-emphasizing practical application of that theory. The rules governing the composition of their thesis committees reflect the need to have experts trained in the theoretical aspects of their field guide them in that process. The involvement of ordinary citizens in research challenges conventional notions about who can produce legitimate knowledge. Even the rules governing the oversight of research by institutional review boards[2] typically deny involvement of community members in developing research protocols and measures for reducing any risks they may face in participating in the research (Bradley, 2007). Ordinary citizens are thus structurally barred from formally participating in the design of research. In addition, students may encounter resistance from their academic committee members to involving ordinary citizens in any aspect of their research. CFERP fellows have encountered such resistance.

The career incentives in the academy also discourage the practice of participatory research. These incentives include greater rewards (honors, awards, promotions, etc.) for contributions to theory rather than practical action, basic rather than applied research and publications in prestigious journals. These are powerful incentives that mitigate against adopting participatory approaches to research. They encourage professional researchers (faculty, graduate students and others) to focus their energies on contributing to scholarship in accepted ways. This creates a tension in the practice of participatory research between the goals and expectations of community members and the goals and expectations of the professional researchers. While community members may look for analytical closure once their problem-solving or political goals have been met, professional researchers will want to continue to explore contradictions and alternative explanations in order to contribute to general knowledge about the topic under study. This questioning and exploration sometimes produces explanations that community members do not like (Firehock, 2003), or that create contradictions between theory and practical action.

An example of these kinds of contradictions in research intended to benefit communities is the recovery of lost knowledge in Native American communities. In their contribution to this volume (see Chapter 11), Hankins and Ross suggest that both native communities and the academy stand to benefit from the recovery of lost knowledge, but that disseminating the knowledge may, in fact, be detrimental to the community. They point out that there is a history of research extracting knowledge from Native American communities and disseminating it in ways that decreases the control that communities have over their own knowledge, traditions and now, with contemporary collection of genetic material, even their bodies. As an example, they cite the frequent disclosure of the locations of sacred sites, which opens those sites to infringement by other researchers and curiosity seekers. They suggest that open communication is needed between the researcher and community members to guard against such outcomes.

Even when open communication is achieved, however, scholars claim some authority over the subject matter through the very act of writing. Indeed, they must in the current educational system, which values single authorship more highly than collaborative authorship. In their writing, scholars are bound to analyze contradictions in their data and/or in the social dynamics of the situation they are studying. These written interpretations, once published, take on an aura of 'truth' that may actually run contrary to the interests of community members in the long run. Many CFERP fellows and researchers in the NCFC's regional centers have wrestled with these questions of what and how much information to publish, and the implications those choices have for rigorous analysis, as well as community empowerment.

This is an issue common to research with communities the world over. Hale (2006), for example, describes an indigenous land claims case being heard in a court in Latin America, and explains that in this case the indigenous people and their professional supporters swayed the court to rule in their favor by arguing that the people have a cultural connection to the land stemming from time immemorial. Hale (2006) observes that although this argument won their case for them this time, such arguments ossify traditional cultures and evoke the very notions of tradition-bound peoples incapable of modernizing, which colonial authorities and even more recent government administrations used to justify assimilation and even genocide in the past. Hale (2006) notes that activist researchers (researchers who use their research in support of community causes) are obligated by their profession to use objective social science to analyze such contradictions, and are simultaneously responsible for contributing to the betterment of the lives of their community partners. He concludes that meeting these dual obligations requires continuously balancing contradictions in our practice, as well as in the knowledge that we produce.

Petras and Porpora (1993) have argued that these often competing obligations in participatory research challenge the very identity of academic researchers. They observe that the traditional identity of academic sociologists 'represents a call to the development of disciplinary theory, methods and substance. Thus, our research is primarily oriented toward the academy, where our findings are evaluated as a contribution to the intellectual community of which we are a part' (Petras and

Porpora, 1993, p120). They suggest that in asking academics to surrender the research agenda to communities and to put practical community benefits on the same level as explaining the world better asks academics to redefine who they are. They argue that 'these forms of participatory research, therefore, create a tension between theory and practice that, for such research to be feasible, needs to be resolved' (Petras and Porpora, 1993, p121). Like Hale (2006), they suggest that this tension must be resolved in a way that retains the benefits to both academic disciplines and communities.

The problem becomes even more complex when there are multiple communities with competing interests. CFERP fellow, Sara Jo Breslow (2006), for example, faced such a situation in conducting research on salmon habitat restoration in Washington's Skagit Valley. The river's delta was diked and drained during the late 19th century, bringing some of the world's most productive arable soils into commercial agricultural production. Current efforts to restore salmon populations in the river to their historic levels include proposals to reforest wide buffer zones along fish-bearing streams. Such a move, while helping fishermen, including Native American fishermen, would harm farmers.

Breslow's (2006) goal in conducting her research was to build an understanding of the habitat restoration issue from the perspectives of all the players involved: fisheries biologists, tribal and non-tribal fishermen, farmers, restoration advocates and government officials. To do so, she chose to 'maintain an aura of neutrality' in order to gain access to multiple communities, rather than to immerse herself in one single community.

This choice had clear implications for community participation in the research. Because of it, she did not establish a close working relationship with any of the communities, although she did achieve some level of trust with individual members of each. She reports that her relationship was best with the fisheries biologists, perhaps because they shared a background in the academy. She also reports that restoration advocates were disappointed and confused when she clearly indicated that her goal was not necessarily to support their conservation efforts (Breslow, 2006). Given the circumstances, Breslow's choice was appropriate for achieving her goal of gaining a broader perspective that could perhaps contribute to a compromise solution between the many communities involved in the conflict. The key point for our argument here, however, is that how one navigates the tension between academic and community interests, or the interests of multiple communities, leads to various trade-offs, with unique implications for the research as well as for community capacity-building.

Breslow (2006) was not alone in choosing a course that precluded some aspects of a participatory design for her research. A tenth-year review of CFERP, completed in July 2007, consisting of in-person visits to several field sites where CFERP has supported research, face-to-face and telephone interviews, review of fellow reports and other documents, as well as an on-line survey, found that CFERP fellows often do not fully engage community members in every aspect of their research.

Table 3.1 *Key ways in which community members contributed to fellows' research (as reported by the fellows)*

	Percentage contributing (n = 37)
Provided data	80
Facilitated introductions	77
Helped to select topic	63
Held meetings	54
Shared results with the community	51
Helped to ensure support of community leadership	51
Helped to design the research	46
Contributed to other products	29
Collected or helped to collect data	29
Continued the work or helped to continue work after student left	23
Reviewed publications	20
Helped to analyze data	11
Contributed to publications	9

As Table 3.1 indicates, community members have tended to be most involved in selecting the topic to be researched, as well as in providing data. Slightly less than half of the fellows responding to the survey engaged their community partners in designing the research, and even fewer engaged them in collecting data, data analysis and other research tasks.

To some extent the tension between community activism and contributing to the development of substantive knowledge and theory lies in the very nature of academic enquiry. Whether one adopts a conventional or participatory approach to research, the overriding professional goal is creating a better account of the world. Since every participatory research project is different, involving different relationships of power, different economic, social and environmental circumstances, and therefore different potential contributions to substantive knowledge, theory and community development, the tension between community and professional researcher expectations for the research will have to be dealt with in unique ways in every case. Participatory research thus requires a continuous balancing of this tension (Hale, 2006).

The current structure of graduate education and academic career incentives often encourages the resolution of this tension in favor of individual and institutional academic priorities, rather than community priorities and community betterment. This, in fact, is the core issue. Research has tended to benefit people, such as professional researchers, government agencies and business corporations, who are already positioned with access to the resources and networks of power that they need in order to apply results to their advantage. Participatory research is intended to change this dynamic in ways that resolve the tension between theory and practice more equitably between community members and professional researchers, and to achieve greater equity in the application of research results. Yet, achieving greater equity requires trustworthy knowledge, the production of which depends upon rigorous analysis of data and scrutiny of alternative explanations. This may, at times, as in Breslow's (2006) case, require working towards a compromise solution with benefits for multiple interests, or working for the benefit of a single community, such as the indigenous community in Hale's (2006) example.

Even when the intention is to benefit marginalized communities, however, the very act of working within the academic system entails some level of accommodation to career incentives and institutional structures. Whole fields of study, such as ethnic studies and women's studies, and research traditions such as cultural critique in anthropology and political ecology, emerged in response to the need for scholarship that was more responsive to the needs of disadvantaged groups. Researchers within these fields and traditions explore the operation of power and processes of oppression in order to improve the lives of the poor and the oppressed. Yet, these fields and traditions have a mixed record in adopting approaches to research that attempt to alter the traditional power relations of the research itself. Some political ecologists adopt a participatory approach in their research, but by no means all. Cultural critique in anthropology may emphasize participatory research even less. Hale (2006) observes that anthropologists working in this tradition have not attempted to change the relationship between professional researchers and community members in standard research practice. He suggests that they have not been overly successful in navigating the tension between community and researcher goals and expectations in research.

Whole programs intended to change the traditional relations of research may be structured in ways that actually reproduce them. CFERP is a case in point. The major activity of CFERP – providing fellowships to graduate students – functions according to the traditional academic model. The program accepts proposals from graduate students enrolled in degree-granting programs around the country, and evaluates them with respect to the rigor of the research question, the participatory design of the proposed research and potential contributions to the field. These evaluation criteria themselves encompass the tension between community and academic expectations for research. Although the selection committee looks for strong evidence of community involvement in the proposed research, aside from pre-dissertation grants that support initial student contact with communities, there is no mechanism for encouraging community initiation of research projects or student–community collaboration on research question formulation and research design. Instead, students typically follow the traditional academic pattern of developing proposals for research based on readings of the academic literature. Thus, although a goal of CFERP is to increase the institutional presence of PR, its design accommodates the traditional approach to graduate education.

This means that students selected as CFERP fellows typically do not have a pre-existing relationship with the communities with whom they propose to do research.

The ten-year review of the program found that having a pre-existing relationship, or taking the time to establish a strong relationship with the community, was a key factor in determining the extent to which CFERP-supported research provided benefits to the community.[3]

Part of the reason that most students do not have pre-existing relationships with communities is because participatory research is not part of standard research practice and there is little support for it in the current academic system despite the presence of faculty and programs committed to participatory research in scattered departments and institutions around the country. One indication of this lack of support for adopting a participatory approach to research in graduate education today is the responses that CFERP fellows gave to a survey question asking what their university could do to better support participatory action research. Eighty-three per cent of the respondents indicated that their institution could acknowledge the validity of the approach for academic research. About three-quarters of the respondents also felt that their universities could provide courses for students and training for faculty in participatory action research, and provide more resources about this approach to research (see Table 3.2). Faculty advisers who responded to the survey felt similarly. Eighty-three per cent felt that their university could provide more resources about participatory action research, and over half felt that their universities could provide courses for students and training for faculty, and acknowledge the validity of the approach for academic research.

With little support for participatory research overall in the academy, overcoming barriers to its practice takes a great deal of commitment and effort on the part of graduate students. While there are a growing number of faculty at universities and colleges around the country who understand and support PR, there is still a strong need for support of participatory research in the academy. Such support is needed in communities, as well, because the biases of the academy are sometimes reflected in the community. In some communities people have deferred to CFERP fellows assuming that they, as researchers, are the experts. For example, Sara Breslow (2006) reports that people whom she talked to wanted *her* to do the research because they did not have time and because she was a neutral outside observer. Moreover, communities and community members themselves vary in their sophistication. Some communities are well organized and prepared for full participation in research, while others need much more organizing and guidance from professional researchers.

The structural biases that affect research conventions in the academy also affect the research of non-profit organizations, such as the NCFC. Part of the problem that the NCFC encountered in institutionalizing participatory research lay in the lack of integration between academic and non-governmental organization (NGO) researchers. From its very conception, the NCFC was established as an organization solely within the non-profit sector. In awarding the grant that created the NCFC as a part of a non-profit organization, the USDA CSREES simultaneously recognized the legitimacy of participatory research and helped to further its institutionalization.

Table 3.2 *Perceived need for support for participatory research in universities*

What could your university do to better support participatory action research (check all that apply)?		
Response	*Fellows (n = 29) (percentage)*	*Faculty (n = 18) (percentage)*
Acknowledge validity of approach for academic research	83	65
Provide courses for students in participatory action research	76	53
Provide training for faculty in participatory action research	76	53
Support faculty to travel to site of student research	48	65
Provide more resources about participatory action research	72	82
Support coordination with other universities	NA	53
Other	14	6

Note: NA = not applicable.

Awarding the grant to a non-profit was significant because, historically, grants of this nature have been awarded only to land-grant universities. Nevertheless, while doing so was more suited to the goals of the grant (since the primary focus of the center's work was building research capacity in communities with whom they partner), a consequence was that the NCFC was a stand-alone project with few formal ties to academic institutions. This limited opportunities for collaborating on projects, exchanging knowledge, sharing resources and skills, leveraging funds and building mechanisms for expanding the institutional presence of participatory research in the academy, as well as in the non-profit sector.

The only institutional link between the NCFC and academic researchers was the National Advisory Council (NAC) of the NCFC. The National Advisory Council was established to guide and support the NCFC, and to involve academic researchers in it. Members of the NAC were leading scientists and community forestry practitioners. The NAC was supposed to meet once every eight months. However, the NCFC found it difficult to engage a diverse and significant number of leading academic researchers in the NAC. One of the main reasons for this was that academic researchers find it challenging to devote time to a non-profit initiative that has no formal and direct links to their academic work. Thus, as an institutional link with the academic community, the NAC was not as effective as originally hoped.

The separation of research and community development had a hand in limiting academic involvement in the NAC. Academic scholars receive the greatest credit for publications of their own original research. Even if their research is participatory, they earn the most credit for single-authored publications. Although community service is expected of faculty, contributions to community development and non-profit organizations are not valued as highly as peer-reviewed publications. In contrast, non-profit professionals get credit for their relationships with communities and their contributions to on-the-ground improvements to the environment, the community or both. These were certainly the criteria upon which NCFC researchers were evaluated (Virtue, 2004). In the case of the NCFC, then, university-based scholars had knowledge, skills and access to resources that could have benefited the regional centers, and the regional centers had established reputations, community networks and community organizing experience that could have enhanced joint participatory research projects and the development of programs of research. There were, however, few incentives in place to encourage sustained interaction between the two. There was, thus, a structural mismatch between the incentives needed to adopt a participatory approach (which lie with non-profits) and the institutional linkages (which lie within the academy) required to encourage further acceptance of that approach within the academy.

Having few strong links between the NCFC and the academy resulted in the NCFC having little impact outside of the communities with which its four centers partnered. While the NCFC had many successful projects with these communities, sustaining the effort beyond the four-year term of the CSREES grant was unsuccessful. CSREES considered its grant seed money and had anticipated the development of additional projects with new sources of funding, as well as strengthened relations between non-profit and university-based researchers. However, the National Network of Forest Practitioners, host of the NCFC, was not successful in its bid to raise funds for the NCFC from other government agencies and private foundations at the national level. The non-profits that hosted the four regional centers could also not obtain funds for continuing their participatory research work at the regional level. As funding started to dwindle, and with the chances of new funding diminishing, the regional centers operated on a shoe-string budget, limiting their work to offering information and publications based on their earlier participatory research projects. But because they could not fund researchers to work with communities, their links to the communities with whom they worked slowly weakened, and the opportunities for reaching out to new communities diminished.

Internal and external factors contributed to the problem. Internally, part of the problem was a lack of cooperation between the four regional centers (Virtue, 2004). Each center operated as its own distinct unit and developed relationships with communities in its own distinctive way. While this was suitable to addressing community issues in the context of the unique circumstances of each region, it did not encourage development of cross-regional cooperation, which could, in turn, lead to forging links with other non-profit organizations or universities.

An external force that hampered the ability of the NCFC to sustain itself is the funding priorities of foundations and other grant-making organizations. The separation of research and community development into distinct spheres of thought and endeavor translates into separate spheres of funding for each. The traditional funders of research do not fund community development, and usually will not fund PR. For example, when one of the authors of this chapter (Wilmsen) enquired about possible funding for CFERP at the USDA's National Research Initiative, he was told that NRI does not fund participatory research at all. On the other side, traditional funders of community development, for their part, do not fund research. The National Forest Foundation, for example, funds projects that entail on-the-ground stewardship. Other conservation and community development foundations typically only want on-the-ground action as well. Although there are exceptions – the Ford Foundation, the Sociological Initiatives Foundation and the Christensen Fund are three examples of foundations that fund PR, and CSREES made a one-time grant to fund the NCFC – foundations often will not fund projects that combine research and action in conservation and community development. This division of labor within philanthropy ultimately contributed to the demise of the NCFC.

It has also been an ongoing challenge for the CFERP in at least two ways. First, the CFERP has encountered difficulties in expanding its funding base for the same reasons that the NCFC had difficulty raising funds. Second, some CFERP fellows with grants from additional funders reported contradictory requirements of their CFERP and other grants.

EFFECTS OF INSTITUTIONAL BIASES ON RESEARCH PRACTICE

The experience of the NCFC is also instructive about how institutional biases – disincentives in the academy to practice participatory research and the disjunction between academic and non-profit objectives, stemming from the separation of research and community development – lead to gaps in techniques and strategies for conducting PR. An internal conflict in the NCFC over the work of the Pacific West Community Forestry Center (PWCFC) illustrates how inadequate strategies for moving from community organizing to research question development affects prospects for the institutionalization of PR.

The conflict arose over the strategy that the PWCFC adopted for working with immigrant forest workers. After extensive consultation with its advisory council, the PWCFC chose to do research with non-timber forest products harvesters in California, Oregon and Washington. The majority of these workers were recent immigrants, with the major proportion being from Latin America and a significant number being from Southeast Asia. Some were in the US illegally. Many did not speak English. In addition to these factors, the seasonal nature of the work, the large geographic area over which the work was conducted, low wages, the frequent use of labor contractors by employers and a number of other conditions presented special challenges in community organizing and in conducting participatory research among these workers.

The basic approach to conducting research that all NCFC centers adopted entailed selecting partner communities (with the help of the center's advisory council), identifying leadership (known as 'local cooperators') who acted as community liaisons during the research, helping the community identify information needs, developing a collaborative work plan, supporting the community in collecting information and helping the community interpret and use the information. While all communities initially needed help and facilitation in formulating questions and developing research methods, some required more assistance with organizational development than others. This led to a significant debate within the NCFC about the degree to which center researchers should engage in organizational development. At issue was the extent to which professional researchers should take the lead in developing research questions as opposed to engaging in a lengthy process of community organizing to develop the community's capacity to define its own research questions.

This issue was not unique to the NCFC. Park (1993) has argued that while communities need varying degrees of organizing before they can formulate research questions, at some point the professional researchers must take the lead in formulating a question. Maguire (1993) similarly describes a situation in which she spent a year organizing a group of battered women to empower themselves and assist one another. Although she gathered enough data in this process to complete her doctoral dissertation, the women's group never reached the point of formulating a research question.

The debate over this issue in the NCFC was intense with regard to the PWCFC and its participatory research project with underserved Latino forest workers in Shelton (Mason County), Washington. The Latino forest workers in Shelton – mainly brush harvesters – were itinerant workers, moving about from place to place in rhythm with the seasonal work cycle. This group faced issues such as disempowerment, discrimination, maltreatment of undocumented workers, insufficient or no health coverage, low wages and poor work conditions and little freedom to discuss them, limited access to places to gather safely, and limited access to information and translation needs (see Chapters 8 and 9 of this volume for case studies of participatory research with immigrant Latino forest workers). Members of the PWCFC's advisory council suggested a participatory research project to improve their working conditions.

In collaboration with the PWCFC, the Latino group began their participatory research work by focusing on identifying leadership within this mobile community through the Latino Forest Workers Leadership Group (LFWLG) project. Using participatory workshops, the LFWLG project tried to build capacity among emerging leaders to design and implement community-based projects and initiatives. The PWCFC expected the leadership to mobilize the workers to identify the type of research questions that were important for their work in the next stage of the research. However, a significant portion of the PWCFC's resources for the project was spent on the leadership project.

This caused a great deal of debate within the NCFC. The lead principal investigator of the NCFC was critical of the PWCFC's focus on organizing the leadership and lack of a research plan with clearly identified research questions and results because he was concerned about meeting the requirements of the CSREES grant that had created the NCFC. In an attempt to force the PWCFC to produce measurable results, he delayed payment of reimbursements to the PWCFC for expenses that it had incurred for this project.[4]

Others within the NCFC weighed in on both sides of the debate. A researcher from another center mentioned that the PWCFC was trying to build a community where none existed. The National Network of Forest Practitioners' (NNFP's) board of directors, for its part, supported the PWCFC, noting that organizing communities so that they themselves can develop the research questions lies at the core of PR, ensuring that people directly affected by the research and its results actively participate in shaping and guiding it. The PWCFC took the stand that until the group was organized it was difficult to talk about research questions and a plan because this should originate from the group itself as the research evolves. The matter was finally resolved when the PWCFC submitted a research plan developed in partnership with the local cooperator from the LFWLG project. That is to say, leaders of the project developed the research plan rather than the workers themselves. The principal investigator then reimbursed the funds for the project. The research continued and the PWCFC continued to engage Latino forest workers as partners in this project.

This debate was significant because it highlights a need in the current practice of PR: researchers need strategies for making the transition from community organizing to jointly developing research questions, and these strategies then need to be incorporated within standard training for participatory research researchers. The essence of the debate within the NCFC was a tension inherent in the process of jointly developing research questions. On one extreme, professional researchers develop the research question with token or minimal community input. This risks conducting research that, at best, is irrelevant to the community and, at worst, contributes to its further marginalization. At the other extreme is a complete hands-off approach in which no question is considered legitimate unless it is developed solely by community members. This risks working indefinitely on community organizing since, depending upon the circumstances, some communities may never reach the point at which they are comfortable defining a research question. Clearly, what is best in each situation will lie somewhere between these two extremes. At present, however, tools and strategies for addressing this issue are not well developed.

Here it is evident that the limited institutionalization of participatory research affects practice. Had there been greater support for the role of community organizing in participatory research within the sponsoring agency (CSREES), this conflict around the PWCFC might not have arisen. The conflict also suggests that practice affects institutionalization. If there were better developed strategies for integrating community organizing and research question development, they would be a part of institutionalized standard research practice. In other words, the strategies are not part of practice, so they have not been institutionalized.

Conclusion: Challenges to the Institutionalization of Participatory Research

The incorporation of participatory approaches, methods and ideas within the formal and informal structures and practices of research (i.e. the institutionalization, as we have defined it, of PR) is incremental. Participatory research is slowly becoming established in the academy and in professional research organizations through the efforts of programs such as the CFERP and the NCFC, as well as through the work of individual scholars and collaborative organizations scattered across the country. At the same time, however, the current structure and practice of academic and non-profit systems, which are based on a fundamental separation of research and practical action (including community development), largely discourage the adoption of participatory approaches. Because of contradictory pressures that pull researchers in different directions, the experience of CFERP and NCFC includes cases in which the commitment of the individuals involved (professional and community), together with strong institutional support, resulted in participatory research projects with clear benefits for the community, as well as projects where institutional support was insufficient and individual personalities clashed. This indicates how important it is to demarcate the relationships of power, the interests of all the players and the institutional barriers to community capacity-building, as well as to address them or somehow account for them in the design of every participatory research project. The experience compels us to conclude with Bebbington (2004, p281) that participation in research, in community development projects, in planning and in other activities should never again be considered without taking into account the material conditions under which it occurs.

There are several outcomes that may result from the particular complex of power relationships, interests, intentions and choices that come into play in every participatory research project. In some cases, participatory research is implemented in ways that empower communities to initiate change. In other cases, agency officials and/or professional researchers may intentionally co-opt community members into projects in order to further their own ends just as community participation has been co-opted in many instances in purported sustainable economic development. In many cases, the intention may not be so pernicious; but professional researchers may simply succumb to pressures to prioritize academic goals ahead of community development goals. In still other cases, community empowerment principles may be observed; but prevailing relationships of power prevent realization of all or some of the empowerment goals.

The lesson for institutionalizing participatory research thus is that institutional biases in the structure and practice of participatory research must be addressed. If they are not, participatory research risks being regarded as just another method for either forcing conservation or development initiatives on local communities or extracting knowledge and information from them in the name of research, the results of which do not produce tangible benefits for community participants. If this risk is realized, it will perpetuate the alienation of community members from programs and initiatives implemented by government agencies, universities, NGOs or other external entities.

To avoid this risk, research and community development challenges must be addressed simultaneously at the community and institutional levels. Work with communities must be rooted in collaborative relationships that increase inclusion, consultation and mobilization for influencing institutions and processes, and work with institutions must make them more accountable and responsive to communities (Gaventa, 2004). When participatory research is used, it should truly engage community members in such a way that their expectations for the research are met to the extent possible given the community's internal and external relationships of power. While this may or may not entail high levels of participation by community members, it does involve establishing a relationship with community members that enables them to make informed choices about their contributions to the research and how they use the results.

The cases of the NCFC and CFERP hold lessons for building acceptance of, and capacity for, participatory research in communities as well as in the academic and non-profit sectors. On the conceptual level, the rigid divide between community development and research needs to be softened so that researchers can more freely use the two to inform one another. On the practical level, the institutionalization of participatory research will be greatly advanced by creating opportunities for sustained collaboration between universities, non-profit organizations and communities; changing career incentives for students and faculty in the academy; providing more training and resources for students and faculty; establishing more flexibility in expectations about the time needed to achieve results to accommodate the specific circumstances of each community; and creating more funding opportunities for academics, non-profits and communities alike. The importance of such measures is evident in the responses that CFERP fellows gave to the survey question: 'Would you have used PAR if you had not received this grant?' The fact that 51 per cent responded 'maybe' suggests that without the support of programs such as the CFERP, and/or stronger support within their own institutions, many students would not adopt a participatory approach in their research projects.

Communities similarly need greater support than they currently receive in capacity-building and training, as well as availability of funds and other resources for building on the findings of research and/or using those findings on their own terms for their own benefit. The CFERP projects with the greatest community benefits are those in which the fellows have a pre-existing relationship with the community or work to build a strong relationship, train community members, fully engage community members in community development or conservation actions related to their research, and/or help community members continue those actions or initiate new community development projects after they completed their research. For example, one CFERP fellow helped Zuni Pueblo in New Mexico start a small-scale sawmill. After completing his research, the fellow stayed with the project as sawmill manager for a year, during which time he trained tribal members in sawmill management. Ten years later, the sawmill is still going strong under tribal leadership.

The NCFC was also most successful in cases where regional center staff enhanced the capacity of community groups. When regional center staff were able to gain the trust of, and work collaboratively with, the community, the partnerships achieved a great deal (Virtue, 2004). Some of the accomplishments included a significant increase in forest restoration contracts for Las Humanas and the land grant communities in the Manzano Mountains of central New Mexico; plans to drop timber sales in important *matsutake* harvest areas and a commitment from the US Forest Service to consult with harvesters on all future plans relevant to the harvest in Crescent Lake in southern Oregon; improved relations between residents and the harvesters in communities in southern Oregon; combining of Maidu traditional ecological knowledge with Western science to design a monitoring plan to assess specific traditional Maidu land management practices; and forming a landowners' co-operative in the central Blue Ridge Mountains that will be a vertically integrated forest products enterprise.

Perhaps a single lesson for the institutionalization of participatory research that can be drawn from the collective experience of the NCFC and CFERP is that flexibility needs to be built into the structures and practices of educational and community development organizations so that researchers and their community collaborators have the time, the funding, the methodological tools and other resources that they need to assess the conditions under which they are co-conducting their research.

NOTES

1 Formerly the Community Forestry Research Fellowship (CFRF) program. The name was changed in December 2007.
2 Institutional review boards are faculty committees charged with ensuring that any research conducted by faculty or students adequately protects the participants in the study from possible risks that their participation might incur. Federal law requires all universities and colleges that sponsor research to form such committees.
3 As a result of the evaluation report, the CFERP is developing means of involving communities in the initiation and design of research.
4 As fiscal agent of the NCFC, the lead principal investigator controlled the funding for all NCFC projects.

REFERENCES

Bebbington, A. (2004) 'Theorizing participation and institutional change: Ethnography and political economy', in S. Hickey and G. Mohan (eds) *Participation: From Tyranny to Transformation?*, Zed Books, London

Bradley, M. (2007) 'Silenced for their own protection: How the IRB marginalizes those it feigns to protect', *ACME*, vol 6, no 3, pp339–349

Breslow, S. J. (2006) *Salmon Habitat Restoration in the Pacific Northwest: Toward Collaborative Stewardship through Participatory Research*, www.cnr.berkeley.edu/community_forestry/People/Final%20Reports/Breslow%20Final%20Report.pdf, accessed 17 December 2007

Bryan, T. A. (2002) *Identity Alignment: The Role of Social Identity in Transforming a Community-based Conflict*, www.cnr.berkeley.edu/community_forestry/publications/reports_2001/bryan_report.pdf, accessed 14 August 2007

Cooke, B. (2004) 'Rules of thumb for participatory change agents', in S. Hickey and G. Mohan (eds) *Participation: From Tyranny to Transformation?*, Zed Books, London and New York, NY

Firehock, K. (2003) *Protocol and Guidelines for Ethical and Effective Research of Community Based Collaborative Processes*, www.cbcrc.org/documents.html, accessed 7 September 2007

Gaventa, J. (2004) 'Towards participatory governance: Assessing the transformative possibilities', in S. Hickey and G. Mohan (eds) *Participation: From Tyranny to Transformation?*, Zed Books, London

Hale, C. R. (2006) 'Activist research v cultural critique: Indigenous land rights and the contradictions of politically engaged anthropology', *Cultural Anthropology*, vol 21, no 1, pp96–120

Maguire, P. (1993) 'Challenges, contradictions, and celebrations: Attempting participatory research as a doctoral student', in P. Park, M. Brydon-Miller, B. Hall and T. Jackson (eds) *Voices of Change: Participatory Research in the United States and Canada*, Bergin and Garvey, Westport, CT

Park, P. (1993) 'What is participatory research? A theoretical and methodological perspective', in P. Park, M. Brydon-Miller, B. Hall and T. Jackson (eds) *Voices of Change: Participatory Research in the United States and Canada*, Bergin and Garvey, Westport, CT

Petras, E. M. and Porpora, D. V. (1993) 'Participatory research: Three models and an analysis', *The American Sociologist*, vol 24, no 1, pp107–126

Ross, J. (2003) 'Bringing the people back to the land: The Tending and Gathering Garden Project', *Regeneration! Newsletter of the Community Forestry Research Fellowship Program*, pp8–10, www.cnr.berkeley.edu/community_forestry/publications/newsletter/Newsletter_summer_2003.pdf

Sanoff, H. (2000) *Community Participation Methods in Design and Planning*, John Wiley and Sons, New York, NY

Sclove, R., Scammell, M. L. and Holland, B. (1998) *Community-based Research in the United States: An Introductory Reconnaissance*, Loka Institute, Amherst, MA
Virtue, M. (2004) *The National Community Forestry Center: A Program of National Network of Forest Practitioners: External Evaluation*, Manuscript in possession of authors

4

From Environmental Racism to Civic Environmentalism: Using Participation and Nature to Develop Capacity in the Belmont Neighborhood of West Philadelphia

William F. Elmendorf and Michael Rios

INTRODUCTION

Early social ecologists viewed the natural environment as a featureless surface on which social patterns and relationships were distributed. Today, the social importance of diverse urban landscapes and parks, ranging from urban wilderness preserves to trees and plants found in community gardens and streetscapes, is advocated. A growing body of literature revolves around the beneficial and connected relationships between nature and social settings and processes such as interaction. This literature argues that nature is a critical component of personal and community well-being and a stimulus for local activism and democracy. Furthermore, it agrees that empowering people to become involved in the process of landscape and park creation builds community capacity and supports community development.

Increasingly, attention has been given to the relationship between nature and communities of color, in particular, African Americans. Much of this work has centered on either empirical research on use, behaviors and attitudes, or case studies of African Americans organizing around environmental justice issues related to deteriorated parks, air pollution, Brownfield sites and other environmental health concerns. This chapter presents a study of a different sort. To better understand and examine how African American neighborhoods utilize nature as a vehicle to build community, this chapter presents a review of community and community development, a review of African American environmentalism and a case study of local environmental efforts in the Belmont neighborhood of West Philadelphia. Based on participant observations and involvement, the authors present the importance of environmentally based projects in this predominately African American neighborhood and argue for participatory methods of study and assistance that support human interaction and civic environmentalism within a particular community setting.

By civic environmentalism, we mean an approach that brings together elements of environmental stewardship with community capacity-building at the local level.[1] Community capacity can simply be defined as the ability to do work together. An inherent element of civic environmentalism is empowered local participation and the utilization of participatory methods to plan for and achieve local environmental projects and goals. Empowerment is facilitated by the provision of childcare and transportation, food and quality information, training and networking. One approach to achieve civic environmentalism through participatory means is community design, which can be described as a movement or force for change in the creation and management of environments for people. Practitioners of community design identify and solve environmental problems where the issue is some combination of social, economic or political problem in nature (Comerio, 1984).

This exploratory chapter focuses on the concept of using a participatory philosophy, such as civic environmentalism, to build community capacity. It begins with a description and overview of nature's role in community capacity-building and development. The second section discusses the role that nature plays in communities of color, in particular, among African Americans. This is followed by an introduction to participatory research as a method of achieving civic environmentalism. We further explore civic environmentalism through a community design case study of the Belmont neighborhood in West Philadelphia. Environmental improvement projects undertaken by local community groups working in collaboration with faculty and students from Penn State University are introduced. The chapter concludes with suggestions of how the tools and ideas of participatory research can serve as a vehicle to build the capacity of both the environment and the community in urban neighborhoods, while challenging stereotypical depictions of the environmental values and attitudes of communities of color.

Nature's role in community capacity and development

Urban landscapes and parks are highly diverse, ranging from natural open space and greenways, to woodlots, trees and plants found in streetscapes and other residential and business landscapes. It has long been understood that the environmental, health, economic, educational, family and social benefits of urban landscapes and parks support community in many ways (Dwyer et al, 1992; Kuo, 2001; Nowak et al, 2001; Kaplan et al, 2004).

Specific to community development, landscapes and parks provide educational opportunities and opportunities for interaction, especially among youth, such as outdoor classrooms and other natural settings where teachers, children and families can interact (Dwyer et al, 1991; Nowak et al, 2001). Urban landscapes and parks may be the only forests that some people ever experience, which provides a context for the creation of values and ethics that urbanites place on the natural environment as a whole (Nowak et al, 2001). Since youth are the foundation of any community, families and children need supportive and healthy environments to encourage positive behaviors and to provide a respite from the challenges of urban life (Taylor et al, 1998; Wolf, 1998). Urban landscapes and parks provide families with places and activities that are not segregated in terms of class, age or skill level for talking, visiting and loving. Natural landscapes can serve in highly urban areas (e.g. inner-city public housing) to create suitable play spaces for children, places not overcome by concrete and asphalt. Levels of play, access to adults and incidences of creative play are greatly increased in urban landscapes containing trees and grass (Taylor et al, 1998). These landscapes provide healthier places where youth and other people can explore, express and develop their human disposition and potential in a socially responsible manner through play, volunteer work, self-exploration, self-insight and team-building. In studies of family structure, public housing and outdoor spaces, some researchers have discovered that resident households in Chicago public housing that had trees and green space exhibited healthier patterns of children's play, fewer violent crimes and incivilities, were more constructive, used fewer violent methods to deal with family conflict, and exhibited less physical violence with partners (Kuo et al, 1998; Kuo, 2003). In a 2003 study, Kuo notes:

> *The link between trees and a healthier social ecosystem turns out to be surprisingly simple to explain. In residential areas, barren, treeless spaces become no man's land, which discourages resident interaction and invites crime. The presence of trees and maintained landscapes can transform these no man's lands into pleasant, welcoming, well-used spaces that serve to both strengthen ties among residents and deter crime.* (Kuo, 2003, p154)

Green landscapes in inner-city neighborhoods have been purported to decrease levels of graffiti, vandalism and even crime. Contrary to the public safety views of many in law enforcement and public works, the planting and maintenance of trees and grass increased people's sense of safety in inner-city neighborhoods and decreased feelings of fear and anxiety (Kuo et al, 1998).

The natural environment of a community plays many roles, including a significant role in the healthy and successful social lives of people by providing shared and structured symbols. These symbols (e.g. historical buildings and structures, monuments, trees and hills) help to ground people in their everyday lives, and as change occurs, they provide residents with a consistent sense of place and comfort (Hester, 1990). Appleyard (1979, p146) further discusses the importance of symbols in everyday life:

> *A city life is, in a very large part, a life lived through symbols. Caring symbols communicate hospitality, responsiveness, assurance, shelter and comfort. They play the host to welcomed guests and can extend the concept of home to neighborhood and city. Ordinary citizens interpret their environment as evidence of the presence of others and actions, services, livability, aesthetics and as a reflection of self.*

Urban landscapes and parks can be shared and structured symbols, caring and supportive symbols that become part of the identity and character of a place and evoke pride, attract outside attention and stimulate economic activity (US EPA, 2002).

Trees, parks and other components of the natural environment become powerful symbols when they are perceived as being representative of a social group (such as a neighborhood), especially when the social meaning of nature plays an influential role in relation to other functions such as family, home, play, health and equality (Appleyard, 1979). Nature again proves important because the sense of self *in* place is more important than simply a sense of place, and people's relationship with their natural environment helps to build a stronger connection to place (Hester 1990). Alexander et al (1977, p798) pointed out:

> *Trees have a very deep and crucial meaning to human beings. The significance of old trees is archetypical; in our dreams they often stand for the wholeness of personality. The trees people love create special places; places to be in and places to pass through. Trees have the potential to create various kinds of social places.*

As such, the significance of the environment's social role in reinforcing a sense of locality or place plays an influential role along with other community development factors. Natural features help to create and maintain a sense of place – that is, a feeling of identification and belonging that is important to people.

Strong emotional ties can exist between people and their natural environment (Dwyer et al, 1991). Greider and Garkovich (1994) argued that landscapes can be 'the symbolic representation of a collective local history and the essence of a collective self-definition'. Social meaning and intention can heighten in cases of environmental conflict or opportunity and, inversely, environmental conflict and opportunity occur in cases where social meaning is especially critical (Appleyard 1979). Gredier and Garkovich (1994) also discussed the social connection between people and their natural environment. They suggest that landscapes can be:

> *The symbolic representation of a collective local history and the essence of a collective self-definition … that what is important in any consideration of environmental change is the meaning of the change for those cultural groups that have incorporated that aspect of the physical environment into their definition of themselves* (Gredier and Garkovich, 1994, p21).

Taken together, the benefits of nature – increased social interaction, health and wellness, and symbolic and emotional value – are supportive of the process of community and encourage a community's capacity to develop. Moreover, these benefits illustrate the important connections between people and nature even in highly urbanized places, and their value in community development strategies.

Community and community development

The notion of community is a fundamental idea to most people and there are many definitions supporting the concept of community as the fabric of local life. Family, supporting institutions (school, church, healthcare, local government, financial institutions), a shared territory, a healthy environment, a common life, collective action and social interaction are elements found in definitions of community (Hillery, 1955; Willkinson, 1991). Community has been conceptualized as having congruence of service area, psychological identification with locality, supporting institutions, local autonomy in decision-making and strength in the horizontal interaction between residents and institutions (Warren, 1972). Willkinson (1991) described three essential properties of community as:

1. a local ecology or an organization of social life that meets daily needs and allows for adaptation to change;
2. a comprehensive interactional structure, or social whole, that expresses a full round of interests and needs; and
3. a bond of local solidarity represented in people acting together to solve common problems.

Robert Nisbet (1953, p77) illuminated the definition of community in these words:

> *Community is founded on people conceived in their whole rather than in one or another role, taken separately, that they may hold in the social order. It draws its psychological strength from levels of motivation deeper than those of mere volition of interest. Community is a fusion of feeling and thought, of tradition and commitment, of membership and volition. Its archetype, both historically and symbolically, is the family*

> *and in almost every type of genuine community the nomenclature of family is prominent.*

The foundation of community is social interaction: individuals and groups working together in pursuit of commonly held goals (Luloff and Swanson, 1995). Social interaction is what creates the fabric of community; it encourages and allows the emergence and development of community (Hillery, 1955). Community has also been defined as an aggregate of people sharing a common interest in a particular locality, as having a quality of interaction that supports a concept of community as experience reinforced by space (Bender, 1978). Thus, community is not just a place: it is a place-oriented process. In this process the physical characteristics and qualities of place, or environment, continue to be recognized as playing important roles in the interaction, health and capacity of community. The natural environment, or green infrastructure of nature, provides opportunities for community by enhancing human health and capacity, and by providing important places and symbols, projects and community issues (Willkinson, 1979, 1991).

In community development, the quality of social interaction, or the comprehensive network of associations and actions among people, institutions and the physical and natural environments, is critical to empowerment. In the development *of* community it is important to pay attention to the quality of relationships among residents, institutions and environments of a locality. Development of community requires attention to cohesive and integrated social, environmental and economic structures (Kaufman, 1959; Wilkinson, 1991). Contemporary theories of sustainable development are consistent with theories of community development and explicitly recognize the critical interdependence of social, economic and environmental factors (Ahern and Fable, 1988). For example, principles of sustainable development (Walter and Crenshaw, 1992) include:

- protecting, preserving and restoring the natural environment;
- including long-term environmental and social costs and issues in estimates of economic viability;
- supporting local business, products and services;
- developing clustered and mixed-use communities;
- using advanced transportation and communication systems;
- maximizing conservation and development of renewable resources;
- establishing recycling programs and industries;
- supporting broad-based education and participatory government.

In progressive theory, community development is more than enterprise development. Although economic goals are an overriding objective in many community development projects, economic development without development of community can be divisive and exploitative and, therefore, unsustainable. Community development comprises the attempts and successes of residents to strengthen themselves and their community. This work is facilitated by a perspective of interaction where channels of cooperation and communication are empowered and maintained, where human relationships are supported and strengthened, and where a shared concept of improvement is mutually developed. A developed community is both improved and its people empowered. A healthy physical and natural environment supports this type of community development work (Kaufman and Wilkinson, 1967; Wilkinson, 1991).

Some communities are better able to deal with problems and opportunities than others. These communities have been called competent communities (Cottrell, 1983). In competent communities, leaders and residents collaborate effectively in identifying the needs of the community, achieve a working consensus of goals, agree on ways to implement agreed-upon goals, and collaborate effectively on required action. These goals also include a focus and action on environmental issues (Willkenson, 1991). The competent community exhibits the interactive traits of community through a high degree of community capacity or the ability to do work together.

A DEEPER LOOK AT THE NATURAL ENVIRONMENT AND COMMUNITY DEVELOPMENT

Early social ecologists viewed the natural environment as a featureless surface on which social patterns and relationships distributed themselves (Firey, 1947). Today, there are substantially different ideas about the relationship of the natural environment to social settings and processes (Wilkinson, 1979, 1991; Nowak et al, 2001). Many authors suggest that ecological well-being is a critical component of both community and individual well-being. Wilkinson (1991, p75) discussed this thought:

> *Social and individual well-being cannot be achieved except in ways that also promote ecological well-being. Ecological well-being, which in a literal sense means the well-being of the house of civilization, refers explicitly to natural and other conditions that support and sustain human life. It is not accurate or appropriate to treat the environment as though it was somehow separate from the social life it supports. An active interdependency characterizes the relationship between social life and its surroundings. References to human–environment separation cannot be justified on any grounds today, if they might have been justified heuristically in the distant past.*

Furthermore, Willkinson (1991) described both human and community capacity as growing from an intimate relationship of trust for both self and society. This potential for capacity is suppressed by deficits in meeting primary human needs and social and cultural patterns (housing, education, healthcare, safety, political representation and recreation) that discourage interaction and community work. These 'patterns' also include environmental racism and unhealthy natural environments, or, phrased differently, healthy natural environments that are not freely accessible to all people.

Nature is fundamentally connected to healthy people and places, and people's relationship with nature changes both with economic stagnation and disinvestment, and with growth and development. Today, social ecologists and other authors argue that healthy and accessible natural environments provide opportunities for people to interact and generalize across interest lines. These opportunities help to develop community: places characterized by shared spatial experiences and concern (Wilkinson, 1991). Promoting the development of technology and economy, the interaction of people and healthy environmental surroundings are all crucial for community. There is an active interdependence between these elements which supports successful community life (Ahern and Fable, 1988).

COMMUNITY CAPACITY: A BUILDING BLOCK OF DEVELOPMENT

The process that leads to community development is called capacity-building. Community capacity is simply defined as the strengths and assets of community members, both individually and collectively brought to a cause. It is related to the term organizational capacity, or the ability to do meaningful work. Frank and Smith (1999, p26) described community capacity as:

> *The ways and means needed to do what has to be done. It is broader than simple skills, people and plans. It includes commitment, resources and all that is brought to bear on a process to make it successful.*

They and other community development scholars argue that in marginalized places it is important that capacity be built before community development can take place. Thus, building community capacity, or the ability of the people of a place to work together towards common goals, is one of the first steps towards community development. Moreover, as people's interaction and participation increase, so does the level of development of community (Hillery, 1955; Wilkinson, 1979; Cottrell, 1983; Ayers and Potter, 1989). The planning, implementation and management of urban forests, parks and open space can serve the purpose of building capacity leading to the development of community. First, these settings can more effectively meet people's needs if local residents are actively involved in planning, decision-making and the implementation of environmental improvements. Second, highly participatory environmental projects also promote organization and structure in the most deteriorated neighborhoods by facilitating interaction (and, thus, capacity) through block clubs, neighborhood organizations, church groups and public–private partnerships (McDonough et al, 1991). The degree to which people are involved in docent programs, tree plantings, environmental restoration projects and other environmental volunteer and educational work increases the positive identification with a locale, improves the quality of relationships between people and organizations, and provides opportunities for skills development and networking. This progressive work increases community capacity and helps to develop community (Garkovich, 1982; Rudel, 1989; Lipkis and Lipkis, 1990; Luloff and Swanson, 1995; Maslin et al, 1999).

Many case studies of inner-city initiatives and projects support using the natural environment as a vehicle to build community capacity in the development of community (Shutkin, 2000; Rios, 2002; Hou and Rios, 2003). These descriptions of practical endeavors agree that planning for, maintaining and managing landscapes and parks does build capacity and supports community development. Tree planting and other environmental projects have been repeatedly used by community organizations such as Philadelphia Green, TreesAtlanta, Friends of the San Francisco Urban Forest, Los Angeles TreePeople and New York GreenGorillas, among others, to rebuild the sense and capacity of community and ameliorate the effects of drugs, crime, violence, apathy and despair in often seemingly hopeless neighborhood settings:

> *Planting a tree enables a person to have an immediate, tangible and positive effect on their environment. It fosters community pride and opens channels for individuals to meet their neighbors, tackle community problems and build neighborhood associations.* (Kollin, 1987, p96)

> *Tree planting fosters community spirit and pride, bringing people together for meaningful purpose that can build the bridges and promote the understandings that bring the neighborhood together. The initial efforts of the tree planters compound themselves as others find in the trees a deeper appreciation of the community, as well as natural beauty. It is the beginning of the formation of new values that is the foundation for city-wide transformation. The newly organized group can further push for bike paths, improvements in public transportation and changes to make the area less congested, less polluted and more livable.* (Lipkis and Lipkis, 1990, pVIII)

The simple act of planting a tree, along with more complex projects of environmental restoration, has positive effects on the environmental, economic and social elements of community. These types of actions are especially important in ignored and disenfranchised places, where the battle cry of community capacity is 'celebrate any success'. But how do tree planting and other environmental projects contribute to addressing structural problems that plague low-income communities of color, such as drugs, violence and poverty? The long-term question is whether a critical mass of required community development activities can be completed to move a place forward.

AFRICAN AMERICAN ENVIRONMENTALISM

The analysis of urban landscape and park preferences and behaviors among ethnic populations has received growing attention and Virden and Walker (1999) provided an excellent overview of this literature. Much of the research on African Americans and environmental issues has centered on either empirical studies of use, behavior and attitudes, or case studies of collective action around environmental racism and justice. This section provides an overview of this literature beginning with attitudinal differences between African Americans and whites. This is followed by an explanation of factors that lead to these differences. Discrimination, marginality and differences in social and cultural norms are offered as the primary reasons. This section also includes more recent challenges to this discourse and the assertion that African Americans are less concerned about environmental issues.

Differences between the participation and preferences between African American and white users of urban parks and forests have been documented by previous studies (Washburne, 1978; Gobster and Delgado, 1993; Floyd, 1999; Cordell et al, 2002; Shinew et al, 2004; Elmendorf et al, 2005). In general, these studies found that African Americans were less likely than whites to participate in underdeveloped and remote areas, and in solitary activities such as jogging, walking, hiking, wildlife photography and wildlife observation. African Americans were more likely to prefer group activities and activities involving social interaction, such as team sports, talking and socializing, rather than nature-based or solitary park pastimes. Whites tended to use parks alone or as couples, while African Americans came in larger groups. African Americans expressed a greater fear of nature, a greater desire for urban environments and less satisfaction with parks in their neighborhood than whites. Several studies also reported that African Americans had higher rates of affiliation with voluntary associations of a social, political or religious nature than whites (Floyd et al, 1994; Elmendorf et al, 2005).

Past research on whites and African Americans has also shown that these groups differ from one another in their urban park and forest landscape preferences (Dwyer and Hutchinson, 1990; Virden and Walker, 1999; Gobster, 2002; Elmendorf et al, 2005). In general, these studies found that African Americans preferred parks characterized by manicured and maintained landscapes. African Americans were more likely than whites to perceive natural landscapes as more worrisome and less aesthetically pleasing than developed environments. Furthermore, African American focus groups preferred recreational settings and landscapes that were well lit and supervised. Whites preferred landscape scenes with trees, dense foliage, overgrown vegetation and densely wooded areas, while African Americans cared more about facility and maintenance aspects. Whites also preferred less management and law enforcement settings than did African Americans. African Americans favored more formal landscape designs and greater openness and visibility than whites.

When examining environmental ethics or attitudes, many authors reported that African Americans had negligible concern for, and involvement in, the environment before 1990 (Taylor, 1989; Baugh, 1991). Survey-based studies reported that a lower percentage of African Americans considered themselves sympathetic towards, and active in, environmental issues (Mitchell, 1980); that African Americans were less likely to perceive environmental hazards in the places in which they lived (Hohn, 1976); and that they were less informed, less aware, less interested and less concerned with environmental issues than whites (Kellert, 1984; Taylor, 1989). In contrast, whites were more likely to discuss environmental problems than non-whites and to join and donate money to environmental organizations (Taylor, 1982). Challenging these assertions, authors argued that African Americans had an increasing concern for their environments (Sheppard, 1995), but did not participate in environmental activities to the extent that whites did (Mohai, 1990; Blahna and Toch, 1993).

Other authors discussed three theories historically used to account for ethnic variation in the understanding and use of the natural environment: marginality, ethnicity, or sub-cultural variation and discrimination (Floyd, 1998; Phillip, 2000; Henderson and Ainsworth, 2001). Marginality, introduced by Washburn (1978), is the idea that African Americans occupy a marginal position in society because they are alienated from social and economic opportunities: such realities as education, income, transportation and political representation. Ethnicity is the idea that African Americans continue to develop as a sub-culture, one retaining historic values and traditions distinct from the general white population. Although the ideas of marginality and ethnicity have been criticized (Floyd, 1998), it is clear that many African Americans have different attitudes, behaviors and needs based on day-to-day and cultural realities. Understanding the differences between ethnicity and marginality is important. The theory of marginality implies that ethnic people will use traditional landscapes if provided transportation, education and other resources. The theory of ethnicity implies that we must learn about and understand the cultural values of ethnic groups and incorporate them into useable designs. This implies the importance of using participatory methods in neighborhood settings where there is racial and ethnic diversity.

Discrimination can be described as when the use or perception of nature is affected by perceived, actual or institutional bias. A number of major outbreaks of racial unrest that occurred during the mid to late 1900s in the US were associated with instances of perceived or real discrimination in urban park settings (Shogan and Thomas, 1964; Kornblum, 1983). More recently, the issue of environmental justice and the promotion of social inclusion in the use and enjoyment of urban parks have been discussed (Floyd and Johnson, 2002; Johnston and Shimada, 2004). Racism in environmental organizations and agencies has also been examined as an explanation for a lack of environmental concern and involvement among African Americans (Taylor, 1989; Shabecoff, 1990; Baugh, 1991). In addition, racial incidents that prevented the use of parks and other natural environments have been documented to explain lower levels of environmental concern and activism (Dunlap and Heffernan, 1975; Philpott, 1978). Some authors believe that both historical and actual discrimination are both under-reported and misunderstood (Floyd, 1998, 1999; Gobster, 2002). Gobster (2003) wrote: 'Discrimination is a serious issue in park management that has begun to receive some attention.' Shinew et al (2004) believed that despite an increasing African American middle and upper class, 'African Americans continue to experience overt and symbolic forms of racism and discrimination.'

According to Gobster and Delgado (1993), discrimination decreases levels of satisfaction associated with a park experience by making the individual feel uncomfortable, possibly resulting in antagonistic behavior such as overt anger and violence in extreme circumstances. Others have observed that as perceived discrimination increased, use of public facilities decreased (Floyd et al, 1994) and that the types of discrimination that must be understood include historical, current, perceived, individual, interpersonal, institutional, actual and overt kinds (Floyd, 1999). As such, discrimination is not always the result of overt racism; but it can result from a lack of knowledge and sensitivity towards certain groups of people, inequities in the quality of facilities, programs and funding, and the uneven quality of park facilities, programs and services in areas with high proportions of ethnic users (Gobster and Delgado, 1993).

While discrimination continues to be influential today, both marginality and ethnicity as explanatory factors for the lack of concern for environmental issues among African Americans have been criticized by a number of authors (Floyd, 1998; Phillip, 2000). One common criticism is that the components involved in these theories are intertwined: it is difficult to identify and separate distinct factors of class and culture for each. Another criticism is that other ideas, which are ignored in these theories, are important. Dwyer and Hutchison (1990) discussed the importance of the desire for interracial contact, or avoidance of it. Other variables to consider when exploring African American environmentalism include age, educational attainment, socio-economic status and family structure (Floyd et al, 1994).

Despite these criticisms, a number of different explanations for African American environmentalism revolve around the ideas of marginality, ethnicity and discrimination. One historic explanation for a lack of environmental concern and activity is that African Americans did not fit the socio-economic profile of traditional environmentalists, typically composed of middle- to upper-class whites (Van Liere and Dunlap, 1980; Taylor, 1989). Environmentalism had been viewed as a white middle class concern due to a higher priority among low-income African Americans that centered on social and economic concerns. As stated by one individual in an early study of African American environmentalism: 'Whites can afford to be concerned with output; we are still concerned with input' (Kreger, 1973, p33). In considering a lack of environmental concern and involvement among African American populations, some argued that environmental issues were irrelevant compared to more pressing needs of jobs, housing, safety, healthcare, social unrest, education and other basic social needs (Commoner, 1971; Kreger, 1973; Bullard, 1990; Baugh, 1991).

Another explanation of a lack of local environmental action has to do with the assertion that African Americans lacked political efficacy and failed to recognize advocacy channels vis-à-vis environmental concerns (Taylor, 1989). Associated with this lack of political efficacy were other explanations, including the lack of monetary resources, knowledge of the political system and technical expertise (Taylor, 1989). An additional explanation is that environmental issues portrayed in the popular press are framed in ways that are unattractive to African Americans' economic development or opportunity, ethnic identity and civil rights (Blahna and Toch, 1993).

Others have hypothesized that African Americans lack an interest in environmental issues due to distinct cultural preferences that differ significantly from whites (Washburne, 1978). The historical effects of slavery have also been identified as a factor for a lack of African American environmental concern and activity. Some claim that (through slavery) African Americans associated land and the natural environment as a source of misery and humiliation, not of peace and fulfillment (Meeker et al, 1973).

As evidenced by these aforementioned claims, debates about African American environmentalism continue today.[2] Beginning during the early 1990s, a number of authors noted an increase in African American environmental concern and action driven by a growing awareness that African Americans had been disproportionately burdened by a wide range of environmental hazards and the recognition that disparate enforcement of environmental funding and regulations contributed to neighborhood decline much like housing discrimination, redlining practices and residential segregation (Bullard, 1990; Mohai, 1990; Baugh, 1991). For example, Mohai and Bunyan's study (1999, p475) on African American environmentalism contradicted stereotypes related to environment and race:

> *We found little evidence to support the theoretical explanations that predict [that] African Americans are less concerned about the environment than whites. To the contrary, we found few differences between African Americans and whites.*

Writing during the 1990s, authors stressed the relevancy of environmental racism: the fact that African American and other lower class neighborhoods were targets of environmental hazards such as polluting industries and waste facilities (Bullard, 1990). In this context, racial differences were most striking in the concern about neighborhood environmental problems, where African Americans indicated significantly greater concern then whites (Mohai and Bunyan, 1999). Prominent civil rights organizations began to voice their concerns about the racial dimensions of environmental issues such as environmental pollution from waste and other industries (polluted and dangerous neighborhoods), and neighborhoods disenfranchised from municipal and other environmental funding and services, including those supporting public landscapes and parks (Brulle, 1994; Salazar, 1996). As an indicator of increased activism, national organizations such as the National Urban League, the Congressional Black Caucus and the Southern Christian Leadership Conference addressed environmental concerns in African American communities (Bullard, 1990; Baugh, 1991). This heightened activity is indicative of increases among African Americans concerning environmental issues. In many African American and other communities of color, residents are organizing around environmental concerns that threaten health, economic vitality and other community concerns.

Today, new alliances are forming, characterized by partnerships between community groups, non-profit organizations and other intermediaries operating at local and regional levels (Mowen and Kerstetter, 2006). In many low-income communities of color, much of this work is being facilitated through the organizational structure of community and university partnerships comprised of local citizen groups and university researchers and students in professional programs such as public health, community planning, landscape architecture and architecture. For example, a 1997 survey conducted by the Association of Collegiate Schools of Architecture identified over 100 community design programs, centers and non-profit organizations in the US and Canada.[3] Of the 123 architecture schools that offer a professional degree in North America, over 30 per cent run university-based community design and research centers. Technical assistance, community outreach and advocacy characterize much community design work emanating from university campuses.

Increasingly popular is the use of participatory methods. These methods have emerged as important approaches to citizen participation in guiding, building and, upon completion, evaluating community-based projects. Participatory tools are vital given that many community groups turn to university research and outreach programs due to the lack of capacity and resources of grassroots organizations. University-based centers get involved in projects at the initial conceptual stages and help groups to frame issues and problems, taking into account complex social, economic and political considerations. Participatory designs, reports, maps and other documents can highlight resource disparities, articulate environmental concerns, such as the prevalence of toxic sites in low-income neighborhoods, or help to organize a community in support of neighborhood improvements, such as public parks and recreational facilities (Hou and Rios, 2003).

COMMUNITY ENVIRONMENTALISM AND PARTICIPATORY RESEARCH

Participatory engagement between faculty, students and local residents provides a local means of measuring results against early defined goals and of identifying critical elements within a project to help further a community's agenda. This approach will also 'put less powerful groups at the center of the knowledge creation process (and) move people and their daily experiences of struggle and survival from the margins of epistemology to the center' (Hall, 1992). Shifting from 'expert' to 'local' knowledge creates the possibility for new sites of enquiry and discovery outside traditional academic settings – for both faculty and students alike. However, the collective benefits of work accrued by participatory research projects can only be realized if knowledge is shared between schools *and* communities.

This section reviews methods of participatory action research (PAR) to highlight the roles that researchers can play in facilitating community capacity-building and development in local settings, especially low-income communities of color. After a description of PAR, we provide an overview of several techniques used in participatory planning and design (e.g. charrettes, focus group meetings and key informant interviews) that provide utility in generating research questions that can be mutually beneficial to researchers and communities alike. The section concludes with several key questions that frame environmental issues within the context of other community issues as a strategy to initiate dialogue from a participatory perspective.

Participatory action research is an important approach to citizen participation in the design and planning of community environments: natural and human made. Within professionally oriented fields, efforts in PAR have focused on community development, resource management, organizational decision-making and community health, among others (Whyte, 1989; Reardon, 1993; Chambers, 1994; Wallerstein, 1997). As an alternative to traditional scientific methods of research, PAR is 'a way of creating knowledge that involves learning from investigation and applying what is learned to collective problems through social action' (Park, 1992). Shifting from 'expert' to 'local' knowledge creates the possibility for new sites of enquiry and discovery outside traditional academic methods and settings.

Within community settings, a sophisticated repertoire of participation methods of design has been developed to involve residents in decision-making and action. Specific techniques now include computer simulations, gaming exercises, visioning and a host of feedback instruments, ranging from visual preference surveys to focus groups and citizen polling. In addition, consensus building, conflict resolution and organizational participation have served as tools to combat problems associated with top-down public process (Sanoff, 2000). One of the techniques used frequently in participatory planning and design is a charrette.[4] The charrette technique embodies core aspects of participatory research, including a focus on collaboration and mutual engagement between researchers and community members (Petras and Porpora, 1993). It is increasingly used with community members as a vehicle to envision change in community environments. The charrette process can be viewed as an intense work session that typically lasts for several days to one or two weeks. A team of researchers and practitioners representing a range of disciplinary expertise is typically assembled as the core team, and work with community members in an interactive manner. Working in a collaborative and iterative manner creates a feedback mechanism throughout the course of a charrette and ensures flexibility and adaptability. Some of the basic elements of a charrette include one-on-one and group discussions; environmental mapping; an assessment of environmental, social and economic conditions; exploratory sessions to frame substantive issues and develop strategies for future implementation; and presentation of ideas in a public forum to present ideas and solicit feedback. As a technique of PAR, charrettes can aid researchers and communities to define problems and identify strategies through collaborative dialogue and visualization. A charrette's value also lies in the use of visual communication and representation through map-making, drawings and multi-media. This is critical to citizen involvement in research where participants speak different languages and have varying educational backgrounds.

The charrette technique mirrors other tools found in qualitative enquiry, including focus group discussions and key informant interviews. Whether used as part of the charrette technique or alone, focus groups are a viable way of gathering information from and including urban and other hard-to-reach populations who do not respond to mail or telephone surveys. There are many publications on focus groups or facilitated group discussions (Elmendorf and Luloff, 2001). In general, focus groups are another tool to gather both diverse and in-depth information about a small number of subjects. As the name suggests, a focus group is an informal discussion in which eight to ten people brainstorm and talk about a topic in their own language and voice with some guidance from a skilled moderator. However, there continues to be skepticism about the use of focus groups, especially from researchers who rely on more traditional research methods. Most of these concerns are largely related to statistical issues of representativeness, the ability to generalize, sample size and the fact that they do not accommodate tests of significance. These arguments miss the entire point of the use of focus groups. Focus groups are used to determine the salience of particular topics to a local audience, understand the language that people employ to comprehend and describe problems and opportunities, and provide valuable information for more harmonious and successful enquiry and decision-making. Focus groups should be conducted until no new information is collected and the researcher and their partners are reasonably certain that a fairly comprehensive account of the issues and problems has been compiled (Elmendorf and Luloff, 2001).

A third technique is the key informant interview. Key informants are spokespeople who, because of their participation in, and knowledge of, a place are asked to describe the events, actions and beliefs, as well as their attitudes towards them (Jacob et al, 1977; Elmendorf and Luloff, 2001). With the assistance of local people, a structured face-to-face interview based on a formal interview schedule is designed and used both to order the basic questions asked and to ensure that the same questions about who, what, when, why and how are used across interviews. As in focus groups, key informant interviews are held until no new information is collected.

Currently, funding agencies, institutions and others working with, or considering working with, inner-city neighborhoods can be very unaware of the needs, uses and attitudes of residents regarding their natural environment. Simply stated, the attitudes of residents in these places toward their natural environment and environmental enhancement projects are not understood, they are only assumed. This type of exclusive planning, which does not recognize or include the realities of the locality, has proven ineffective many times in the past. It is important to note that techniques such as charrettes, focus groups and key informant interviews are different ways of collecting data in PAR projects. From this perspective, the aim of these techniques is to generate questions *with* community participants at the beginning of research projects – the essence of participatory research – rather than to validate questions, answers and outcomes already formulated by the researcher. Within the field of community forestry, McDonough et al (1991, p75) illustrate this point:

> *Social forestry can be more effective in meeting human needs if communities are integrally involved in the planning, decision-making and implementation of initiatives. Social forestry projects are highly participatory. The increased sense of ownership from such community-based participation leads to improved rates of success in contrast to projects that are directed with a top-down approach.*

As discussed in this chapter, a wide array of recent research has demonstrated that the particular needs, behaviors and attitudes of inner-city African Americans and other ethnic groups towards trees, parks and other natural resources warrant attention. The use of PAR techniques is one way of addressing these questions and of overcoming the selfish divides that often characterize researcher and researched relations.

In practical terms, a number of questions became apparent to consider during charrettes and in focus groups and key informant interviews. These questions may allow outside groups providing funding and assistance to better understand the relationship between local cultural norms, social values and the natural environment. These questions may also give an opportunity for community members to voice their concerns and substantiate the realities of discrimination, marginality and ethnicity in African American and low-income communities of color. Several key questions include the following:

- What are the most pressing problems or concerns in the community?
- Is the natural environment (e.g. trees, parks and open space) important and, if so, why?
- What are the barriers to the use of parks, open space and other recreational facilities?
- Has the community received adequate funding and services for parks and public landscapes?
- Can tree planting and other environmental projects help the community; i f so, how?
- How could the natural environment of the neighborhood change for the better?
- Can neighborhood groups be successful in completing tree planting and other environmental projects; if so, how?

Some readers may question the phrasing and language of the above questions. The value of these questions relies less on their statistical meaning than on their means of engaging in meaningful dialogue between researchers and community members. Questions such as these also create a space for participants to make specific claims about the use and meaning of environments that are the settings for their everyday lives. Semi-structured and open questions, active listening and dialogue are primary tools of participatory research. Given the philosophy of participatory research, any final questions, including their language, should be framed by the understanding, input and agreement of local people who are involved in information gathering.

The next section presents a case study of the Belmont neighborhood of West Philadelphia, where environmental issues and opportunities are entangled with extreme poverty and the hopes, desires and everyday needs of local residents. The case illustrates the value of participatory methods to frame issues, identify mutual research agendas and implement collaborative projects. More generally, the case supports a broader definition of nature – a juxtaposition of the natural environment with the built and social environments – calling for researchers and agencies to acknowledge the important interrelationships between the natural and human-made landscapes which unique people inhabit.

CASE STUDY: THE BELMONT NEIGHBORHOOD IN WEST PHILADELPHIA

Neighborhood overview

The Belmont neighborhood is located in West Philadelphia (see Figure 4.1). While the City of Philadelphia has lost 25 per cent of its population over the last 50 years, West Philadelphia lost 50 per cent during that same time period.[5] The Belmont neighborhood has experienced a drastic decline in total population since

Source: School of Architecture and Landscape Design, Penn State University

Figure 4.1 *The Belmont neighborhood is located in West Philadelphia, northwest of the University of Pennsylvania*

1950, with the greatest decline occurring during the 1970s. Over 95 per cent of the households in Belmont are African American and single unmarried mothers with young children head a majority of these households. There is also a noticeable concentration of seniors living in certain parts of the neighborhood.

As in many American inner-city neighborhoods, Belmont faces serious social and economic problems in some ways depicted by its built environment, with deteriorating structures, vacant lots, declining homeownership and few safe playgrounds. Homeownership, a good index of neighborhood stability, has declined from 56 per cent in 1990 to 49 per cent in 2000 as compared to West Philadelphia (52 per cent to 50 per cent) and Philadelphia (62 per cent to 59 per cent).[6] More than half of all vacant structures in West Philadelphia are in two adjacent neighborhoods: Belmont and Mantua. In 2000, the vacancy rates of housing in Belmont stood at 22 per cent, double the rate of the City of Philadelphia.[7] The total number of housing units in the Belmont neighborhood continues to decline. During the 1990s, Belmont alone lost over 200 housing units to demolition.

The landscape of Belmont's streets presents stark contrasts. Well-maintained houses with neatly trimmed hedges and porches with potted plants stand next to burnt-out structures that are literally falling apart. There are many rubble-strewn vacant lots; but not all are that way. Some residents have organized and fought back to reclaim these spaces by growing vegetable and flower gardens. In fact, the group of neighborhoods north of Market Street and East of 52nd Street, to which Belmont belongs, contains more private and community gardens than any comparable area of the entire city of Philadelphia (Hu, 2003). Contrasts in the physical landscape mirror what exists in the social world of Belmont: hope, love, kindness and faith reside side-by-side with drug addiction, despair, cynicism, simmering anger and crime:

> *We have learned to depend upon each other to get things done. I know the names of most people in this community and in a lot of ways they are like family to me. The children are important and through our after-school program and summer camp, we hope to build a new generation of leadership. We are also proud of our women who live in the single room occupancy hotel. Many of them found ways to turn their lives around and have become leaders within and outside the community. Of course, this is a rough neighborhood to live in at times; but we're here today and we're going to be here tomorrow. We will not let the bad elements drive us out because this is our neighborhood.*[8]

While there is reason for optimism, there are barriers to the use and enjoyment of the natural environment in Belmont. As is the case in many inner-city neighborhoods, drugs and crime have created a psychology of survival in the streets of Belmont. A dialogue with residents illustrates this point:

> *Life is hard in Belmont, or West Philadelphia for that matter. Many of us do not come from 'down the bottom' [slang for Belmont]; but we found our way here from other parts of West Philly to live in the single-room occupancy [SRO] hotels. The SRO has been good to me and I like living there. You ask whether it is hard for me to go out into the neighborhood or do I feel safe? I go just about anywhere I please and nobody is going to stop me, except bad days. Then drugs are all over the neighborhood and sometimes it is too much for me to face. Some of the other women are afraid to go out, especially at night, you know, because of the 'activities' [neighborhood drug dealing]. One of the ladies in the SRO actually walks down the center of Preston Street when she comes home at night. She knows that there's a sidewalk but she says that they aren't lit very well and it feels safer walking down the middle of the street. Did you know that you can see two drug houses from my bedroom window in the SRO? It don't matter how many times the cops shut down these drug houses, those dealers are like cockroaches. They're not going anywhere. But it would be nice to walk down the street and feel safe.*

Physical and mental barriers such as crime, fear of crime and pedestrian safety limit access to parks and playgrounds, the primary vehicle for neighborhood recreation and exercise. In fact, Belmont has no parks or playgrounds. Most of the shade trees that once lined Belmont's street are now gone due to elimination and lack of maintenance by the City of Philadelphia.[9] Despite being in walking distance from Philadelphia's Fairmount Park (the largest urban park in the US), few adults and children of Belmont go there. The area between Fairmount Park and Belmont is quite uninviting – dominated by a railroad, a wide street, streetcar lines and unsafe streets with rows of abandoned houses.

The cumulative effects of a deteriorating neighborhood and unhealthy and inaccessible natural environments can have serious health consequences (Estabrooks et al, 2003; Pettit et al, 2003). The principal health problems of Belmont and West Philadelphia have been listed as high blood pressure, obesity, diabetes, heart failure, back pain and asthma.[10] Many of these ailments have their origin in poor diet and lack of exercise. In the zip code in which Belmont is located, 21 per cent of the adults rate their health status as fair or poor. In addition, 17 per cent of adults reported elevated cholesterol levels, 29 per cent hypertension and 17 per cent obesity.[11] These percentages exceed targeted percentages for federal health goals.[12]

Building capacity through collaboration

While these statistics are shocking, beginning in 2002, Belmont organizations and residents began working together with Penn State University faculty and students to address their social, economic and physical well-being through a series of environmentally focused community development projects. Initially, residents and community leaders challenged the faculty to make a more permanent commitment to the neighborhood and move from providing student service projects lasting only one month in duration to a multi-year initiative that would address pressing neighborhood needs and build community capacity over time through a deeper and stronger relationship. This initiative, the Belmont Community University Partnership (BelCUP), emerged out of the work of Dr Lakshman Yapa, a geography professor at Penn State. Since 1998, Dr Yapa has coordinated the Philadelphia Field Project. Every summer, a select group of undergraduates from a variety of disciplines undertake research-based thesis projects designed to advance understanding of specific aspects of urban poverty. The students complete work in the spring semester and spend a month living, working and conducting research in Belmont. Students who participate in the Philadelphia Field Project conduct research on problems that community members wish to see solved.

The expanded collaboration between Penn State and the Belmont community was facilitated through the Hamer Center for Community Design, a research and outreach unit of the School of Architecture and Landscape Architecture.[13] Faculty and students organized a series of projects resulting in the implementation of small-scale improvement projects and the formation of a nascent stewardship group to conduct assessments, and to plan and implement neighborhood environmental projects (see Figure 4.2). Providing organizational support to BelCUP, the Hamer Center set up a satellite office in West Philadelphia through Penn State's Cooperative Extension and secured a three-year Americorps VISTA grant to assist with project coordination and community outreach. These and other efforts helped to expand Penn State's involvement in Belmont, including participation by the Committee for Community Directed Research and Education (CCDRE). As a faculty and student initiative of the adult education program at Penn State, CCDRE's mission is 'to promote research as service that enables members of communities to define and address their own concerns'. This is carried out through participatory research projects. CCDRE serves as a clearinghouse to assist in community-directed projects and facilitates opportunities for student service learning and internships. The educational emphasis of CCDRE brought an evaluative component to the research, outreach and teaching activities in the Belmont neighborhood.

Source: The Penn State Hamer Center for Community Design Assistance

Figure 4.2 *Local youth assist Penn State students in the building of the Holly Street community garden in 2004*

Participatory practice in action

Of the different research and outreach units of Penn State involved in the Belmont neighborhood, the Hamer Center's activities centered on environmental stewardship through PR. This approach is defined by a commitment to building capacity and providing technical assistance to low- and moderate-income communities working to improve social, economic and environmental health through participatory design and civic environmentalism methods. A core value of community-based design and environmentalism is active engagement by locally vested groups and individuals, and is understood as a critical component in building capacity for community decision-making, the implementation of local programs and successful outcomes on the ground (Kretzman and McKnight, 1993).

The charrette technique was one of the means used by the Hamer Center to identify potential civic environmental projects in the Belmont neighborhood (see Figure 4.3). Working in collaboration with faculty and students from Penn State University over a two-year time period, some of these projects have been implemented by residents, while others have served as the basis for subsequent discussion and research. The Hamer Center organized the first community charrette in 2002, in collaboration with the Friends Rehabilitation Program (FRP), a non-profit Quaker community development corporation working in the neighborhood. Since 1988, FRP, in cooperation with the Belmont Improvement Association, had built or renovated over 100 dwellings. The purpose of the organization is to 'provide affordable housing, community space and service to the elderly, handicapped, low- and moderate-income families and area homeless persons'.

Deborah Thompson-Savage, then a staff member for FRP and a local Belmont resident, invited the Hamer Center to facilitate the initial charrette. During the five-day visit, a team of nine students and two faculty members participated in roundtable discussions with representatives from the City of Philadelphia Planning Commission, FRP and residents of the Sarah Allen community housing facility. The team went on a guided tour of Belmont and spent a day taking pictures, making observations and getting to know the neighborhood. The week's activities culminated with a community workshop that included over 40 neighborhood residents, including representatives from FRP, seniors, women and youth from an after-school program (see Figure 4.4). The roundtable discussions and the community workshop helped to define important neighborhood issues and opportunities, as well as visions for the future of the Belmont Neighborhood. This charrette resulted in ideas for a neighborhood park and facilities to support youth, women and senior citizens living around the Sarah Allen Lucretia Mott Community housing facility in Belmont, owned and maintained by the FRP. While the charrette produced tangible projects to be implemented, it also raised key issues regarding the overall environmental quality of the neighborhood and its relationship to health, safety and accessibility.

Source: The Penn State Hamer Center for Community Design Assistance

Figure 4.3 *Students worked in teams to develop maps and analyze data during one of several charrettes in 2002 and 2004*

After the initial charrette in January 2002, a second one was organized and several environmental improvement projects emerged as a focus for community-building activities. These included a new neighborhood park called the Belmont Commons that would provide much needed green space in the community, identification of vacant lots as potential sites for community gardens and parks, and streetscape improvements, such as lighting and tree planting along major streets to create safer walking environments between neighborhood destinations (see Figure 4.5). These projects were substantiated by a neighborhood inventory that revealed a significant lack of existing trees, parks, open space and lighting.[14] Improvements to two community gardens and land acquisition for the Belmont Commons have already taken place (see Figure 4.6). Additionally, several plans have been completed, including a feasibility plan for the Belmont Commons and a streetscape improvement plan for portions of Lancaster Avenue, the commercial corridor that serves the Belmont Neighborhood.[15]

What is important to point out about these collective activities is that ideas for environmental improvements emerged from community residents, leaders, youth and seniors. The role of faculty and student participants was to, both

Source: The Penn State Hamer Center for Community Design Assistance

Figure 4.4 *Local youth participate in a gaming exercise to envision environmental change in the neighborhood during a community charrette in 2002*

literally and figuratively, 'draw' out hopes, needs and visions articulated by Belmont neighborhood members. Initially, students had preconceptions of the Belmont neighborhood, but soon realized that environmental improvement and community development are complementary activities:

> *I am from the suburbs and we have our share of problems; but I could not believe the conditions people live in here. I've never seen so*

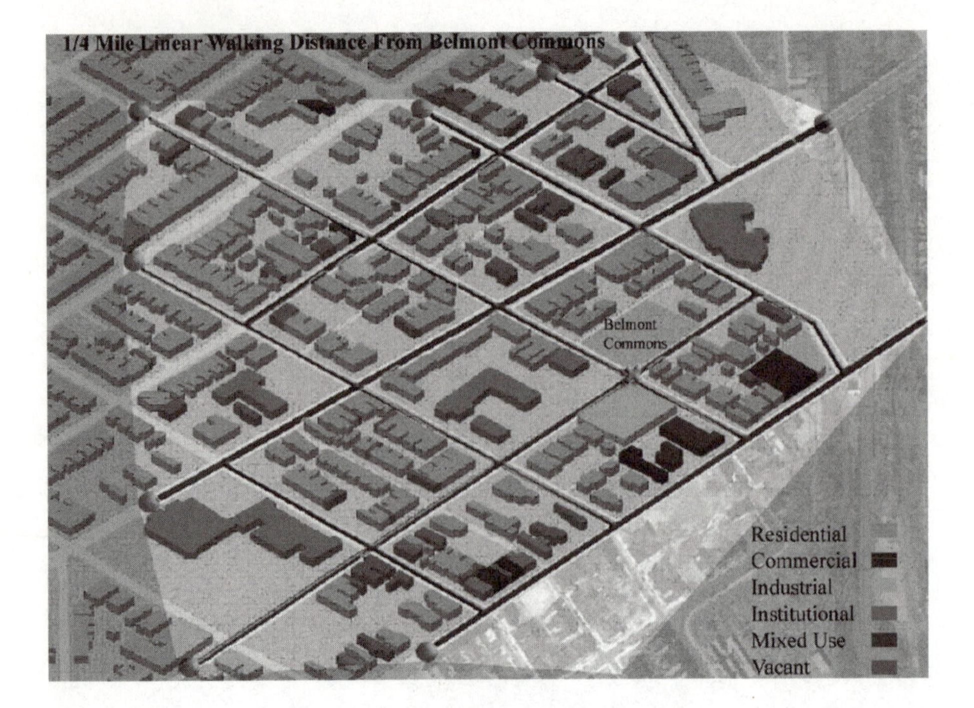

Source: School of Architecture and Landscape Design, Penn State University

Figure 4.5 *GIS maps generated from the neighborhood inventory helped to identify sites for environmental improvement projects and the adjacency of these sites to other community amenities*

> *many abandoned houses and trash filled lots. While I did appreciate the historic character of the old brick row houses and I could see their potential, I initially questioned what could possibly be done to help this neighborhood? After the first day of talking to community members it was amazing to see the positive outlook of many individuals. I learned what could be accomplished through cooperation and persistence. Neighborhoods cannot be physically rebuilt without paying attention to building people as well. This neighborhood is really a lot like where I come from; the children here love to play and have hopes and dreams and most of the people simply live one day at a time, making the best out of what they have.*

Most of the improvements suggested by the charrette participants centered on quality of life issues and the role that environmental improvement projects could play as a community capacity building activity for youth, seniors and the neighborhood.

Source: The Penn State Hamer Center for Community Design Assistance

Figure 4.6 *Penn State architecture students built community garden structures as part of several environmental improvement projects during the spring of 2004*

While the tree planting and pocket park projects described in the case of Belmont can be considered small in comparison to many economic, social and environmental improvements important in community development, collectively, these efforts have begun to establish trust and interaction between local people and local and outside organizations. These projects are helping to develop the ability of community groups to undertake increasingly complex projects, which is the definition of community capacity. In particular, the piqued interest in environmental projects had created the desire to formalize relationships between local organizations such as FRP and the Holly Street Community Garden and Literary Association, a local community organization that uses community gardening and horticulture to teach life skills to neighborhood youth. A relationship was also formed with the Center to Advance Population Health, a community health research and outreach unit at Thomas Jefferson University in Philadelphia that included targeted health services to Belmont residents. The reality of the environmental projects in places such as the Belmont neighborhood is that they are part of a larger social ecology of interdependent human and environmental relations. Without an integrative approach, the likelihood of successfully implementing environmental and natural resource projects is slim. In sum, participatory research and technical assistance projects are only the beginning of a longer process of community development. As such, they should be viewed as vehicles of both environmental stewardship and capacity-building, which help in community development.

CONCLUSION

From social, human health and economic standpoints, tree planting, urban gardening and other collaboratively planned and completed environmental projects are some of the simplest, most rewarding and most celebrated actions that can be used in places such as the Belmont neighborhood of West Philadelphia. It is clear that accessible, high-quality environments and place-oriented environmental projects help to increase the overall quality of a neighborhood's physical environment and capacity. Such efforts are a powerful community development tool. However, their use alone will not overcome the immense problems when considering those places faced with drugs, anger, violence and poverty.

The relationships between community institutions are fragile, especially in inner-city settings. Staff turnover, competition for limited city resources and changing organizational priorities are some of the problems that continue to challenge groups such as FRP and the Holly Street Community Garden and Literary Association. For participatory research efforts, such as those organized by the Hamer Center, a fundamental question is: how much capacity-building support is enough before communities must succeed or fail on their own? Given that the Hamer Center is no longer involved in the Belmont neighborhood, it is now up to the various community organizations and residents to work together to secure the necessary financial and organizational resources to implement priority projects such as the Belmont Commons.[16]

Individuals interested in developing community are concerned about the quality and capacity of relationships among residents, economic advancement and the quality of the physical and natural environment. In 1998, Kenneth Wilkinson described a number of ideas that are important to people who want to improve their community. He posited that community development is about the development of a human relationship structure, not just about things; community development must reflect and express the values and wishes of the local population; community development requires that attention be given to all areas of local life; community requires interaction and interaction requires trust, communication and alignment; and community development requires a commitment beyond selfish gain. Not surprisingly, these ideas in many ways reflect the core philosophies of participatory research and its often overlooked connection to development *of* community.

The case study of the Belmont neighborhood of West Philadelphia was presented to highlight the value that African American communities place on environmental issues and projects. It also demonstrates the difficult social situation found in many inner-city neighborhoods and how the utilization of a participatory approach to project planning and research can be mutually beneficial to both community collaborators and researchers. Additionally, it also challenges critics of participatory community-based resource management who define people and environments too narrowly or incorrectly without considering the full range of benefits of community-based models (Bradshaw, 2003). It is clear that research, environmental projects and landscapes imposed on local people by outsiders mean and do little. What benefits are achieved if stereotypes, misunderstandings and prejudices are continually perpetuated by researchers and professionals who alone identify causal relationships, on the one hand, and expert solutions on the other, without truly engaging in any meaningful dialogue *with* community members?

Participation and understanding takes work: cultivation, time, listening and flexibility. Engaging peoples' participation can be difficult even when using techniques such as charrettes, focus groups, key informant interviews or other PAR methods. We believe that listening to and involving historically disenfranchised populations should not be ignored, but rather recognized as a vital component of any research project where these populations live. Listening to the deep knowledge of local communities provides a dialogical space for better understanding and successful place-oriented action that increases the capacity of *all* people.

NOTES

1 There is a growing literature about civic environmentalism focusing on policy considerations and participation by local communities (see Shutkin, 2000, and Sirianni and Friedland, 2001). However, this discussion is not the focus of this chapter.
2 The post-modern and post-structuralist turns in the academy may explain some of the reasons why a shift has resulted in how African Americans are (re)presented in this literature. Causal explanations of race and ethnicity have been largely dismissed by many social theorists and qualitative researchers that take a social constructivist view. Central to social constructivism is the process by which people (and researchers) in a particular setting construct reality and how values, beliefs and worldviews shape attitudes and behaviors.
3 J. Corey, *ACSA Sourcebook of Community Design Programs* (ACSA Press, 2000, Washington, DC)
4 The French word *charrette* means 'little cart' and is used to describe the final intense work effort expended by art and architecture students to meet a project deadline. At the Ecole des Beaux Arts in Paris during the 19th century, proctors circulated with carts to collect final drawings, while the students frantically put finishing touches on their work (see www.charretteinstitute.org/charrette.html).
5 *Lancaster Avenue Corridor Comprehensive Plan 2002* and US Census 2000.
6 US Census 1990 and 2000.
7 US Census 2000.

8 The narratives in this portion of the chapter summarize roundtable discussions and community workshops that have been held in the Belmont neighborhood. They are not specific testimonials of an individual, but rather drawn from several individuals and are written in a style to capture the essence of a variety of viewpoints (Rios et al, 2002).
9 Pers comm with Bettye Ferguson, director of Holly Street Community Garden and Literary Association (September 2003).
10 See http://westphillydata.library.upenn.edu/infoR_WestPhiladelphia.htm#health table.
11 Philadelphia Health Management Corporation (2002) *Community Health Database Household Survey 2002*.
12 See www.healthypeople.gov.
13 As a university-based community design center, the Hamer Center provides outreach and research support aimed at providing local groups with the technical means to implement community-driven projects. The Hamer Center focuses on a variety of collaborative and interdisciplinary activities for faculty and students ranging from urban design and planning to neighborhood revitalization and affordable housing.
14 The inventory resulted in the creation of a GIS database to be used for subsequent planning of environmental improvement projects. The survey protocol for the inventory was designed to enable Belmont residents to participate in data collection. Related activities included several training workshops and field testing of the survey instrument.
15 *Belmont Commons: A Vision for a Neighborhood Park* (Hamer Center, 2005); *Lancaster Avenue Urban Design Study* (Piehl and Zhang, 2003).
16 A feasibility report prepared by the Hamer Center (2005) included a park master plan, construction cost estimates, a parks partnership framework and identification of resources to implement the project.

References

Ahern, J. and Fabel, J. (1988) 'Linking the global with the local: Landscape ecology, carrying capacity and the sustainable development paradigm', *Proceedings of the Landscape/Land Use Planning Committee of the American Society of Landscape Architecture*, American Society of Landscape Architecture, Washington DC

Alexander, C., Ishikawa, S. and Silverstein, M. (1977) *A Pattern Language: Towns, Buildings, Construction*, Oxford University Press, New York, NY

Appleyard, D. (1979) 'The environment as a social symbol within a theory of environmental action and perception', *American Planning Association Journal*, April, pp143–153

Ayers, F. and Potter, H. (1989) 'Attitudes toward community change: A comparison between rural leaders and residents', *Journal of the Community Development Society*, vol 20, no 1, pp1–18

Baugh, J. (1991) 'African–American and the environment: A review essay', *Policy Studies Journal*, vol 19, no 2, pp183–191

Bender, T. (1978) *Community and Social Change in the United States*, Rutgers University Press, New Brunswick, NJ

Blahna, D. and Toch, M. (1993) 'Environmental reporting in ethnic magazines: Implications for incorporating minority concerns', *Journal of Environmental Education*, vol 24 no 2, pp22–29

Bradshaw, B. (2003) 'Questioning the credibility and capacity of community-based resource management', *The Canadian Geographer*, vol 47, no 2, pp137–150

Brulle, R. (1994) 'Discourse and social movement organizations: The creation of environmental organizations in the United States', Paper Presented at 89th Meeting of American Sociological Association, George Washington University, Washington, DC

Bullard, R. (1990) *Dumping in Dixie: Race, Class and Environmental Quality*, Westview Press Boulder, CO

Chambers, R. (1994) 'Participatory rural appraisal (PRA): Challenges, potentials and paradigms', *World Development*, vol 22, no 10, pp57–72

Comerio, M. (1984) 'Community design: Idealism and entrepreneurship', *The Journal of Architecture and Planning Research*, vol 1, pp227–243

Commoner, B. (1971) *The Closing Circle: Man, Nature and Technology*, Alfred Knopf, New York, NY

Cordel, H., Betz, C. and Green, G. (2002) 'Recreation and the environment as cultural dimensions in contemporary American society', *Leisure Science*, vol 24, pp13–41

Corey, J. (2000) *ACSA Sourcebook of Community Design*, ACSA Press, Washington DC

Cottrell, L. (1983) 'The competent community', in R. L. Warren and L. Lyon (eds) *New Perspectives on the American Community*, Dorsey Press, Homewood, IL

Dunlap, R. and Heffernan, W. (1975) 'Outdoor recreation and environmental concern: An empirical examination', *Rural Sociology*, vol 40, spring, pp18–27

Dwyer, J. and Hutchison, R. (1990) 'Outdoor recreation participation and preferences by black and white Chicago households', in J. Vining (ed) *Social Science and Natural Resource Recreation Management*, Westview Press, Boulder, CO

Dwyer, J., Schroeder, H. and Gobster, P. (1991) 'The significance of urban forests: Towards a deeper understanding of values', *Journal of Arboriculture*, vol 17, no 10, pp276–284

Dwyer, J., McPherson, E., Schroder, H. and Rountree, R. (1992) 'Assessing the benefits and costs of the urban forest', *Journal of Arboriculture*, vol 18, pp227–234

Dwyer, J., Nowak, D., Noble, M. and Sisinni, S. (2000) *Connecting People with Ecosystems in the 21st Century: An Assessment of our Nation's Urban Forests*, USDA Forest Service, Washington, DC

Elmendorf, W. and Luloff, A. (2001) 'Using qualitative data collection methods when planning for community forests', *Journal of Arboriculture*, vol 27, no 3, pp139–151

Elmendorf, W., Willits, F., Sasidharan, V. and Godbey, G. (2005) 'Urban park and forest participation and landscape preference: A comparison between blacks and whites in Philadelphia and Atlanta, US', *Journal of Arboriculture*, vol 31, no 6, pp318–326

Estabrooks, P., Lee, R. and Gyuresik, N. (2003) 'Resources for physical activity participation: Does availability and accessibility differ by neighborhood socioeconomic status?', *Annals of Behavioral Medicine*, vol 25, no 2, pp100–104

Firey, W. (1947) *Land Use in Central Boston*, Harvard University Press, Cambridge, MA

Floyd, M. (1998) 'Getting beyond marginality and ethnicity: The challenge for race and ethnic studies in leisure research', *Journal of Leisure Research*, vol 30, pp3–22

Floyd, M. (1999) *Social Science Research Review: Race, Ethnicity and Use of the National Park System*, National Park Service, Washington, DC

Floyd, M. and Johnson, C. (2002) 'Coming to terms with environmental justice in outdoor recreation: A conceptual discussion with research implications', *Leisure Science*, vol 24, pp 59–77

Floyd, M., Shinew, K., McGuire, F. and Noe, F. (1994) 'Race, class and leisure activity preference: Marginality and ethnicity revisited', *Journal of Leisure Research*, vol 26, pp158–173

Frank, F. and Smith, A. (1999) *The Community Development Handbook: A Tool to Guide Community Capacity*, Human Resources Development Canada, Quebec, Canada

Garkovich, L. (1982) 'Land use planning as a response to rapid population growth and community change', *Rural Sociology*, vol 47, no 1, pp47–57

Gobster, P. (2002) 'Managing urban parks for racial and ethnically diverse clientele', *Leisure Science*, vol 24, pp143–159

Gobster, P. and Delgado, A. (1993) 'Ethnicity and recreation use in Chicago's Lincoln Park', in Gobster, P. (ed) *Managing Urban and High-use Recreation Settings*, General Technical Report NC–163, USDA Forest Service North Central Experiment Station, St Paul, MN

Greider, T. and Garkovich, L. (1994) 'Landscapes: The social construction of nature and the environment', *Rural Sociology*, vol 59, no 1, pp1–14

Hall, B. (1992) 'From margins to center? The development and purpose of participatory research', *The American Sociologist*, winter, pp15–28

Henderson, K. and Ainsworth, B. (2001) 'Researching leisure and physical activity of women of color: Issues and emerging questions', *Leisure Science*, vol 23, pp21–34

Hester, R. (1990) 'The sacred structure in small towns: A return to Manteo, North Carolina', *Small Town*, January–February, pp4–21

Hillery, G. (1955) 'Definitions of community: Areas of agreement', *Rural Sociology*, vol 20, no 2, pp111–125

Hohn, C. (1976) 'A human–ecological approach to the reality and perception of air pollution: The Los Angeles Case', *Pacific Sociological Review*, vol 19, January, pp21–44

Hou, J. and Rios, M. (2003) 'Community-driven place making: The social practice of participatory design in the making of Union Point Park', *Journal of Architectural Education*, vol 57, no 1, September, pp19–27

Hu, E. (2003) *Using Computer Mapping for Community Development in Belmont-Mantua*, Honors Thesis, Schreyer Honors College, Pennsylvania State University, University Park, PA

Jacob, S., Bourke, L. and Luloff, A. (1997) 'Rural community stress, distress and well-being in Pennsylvania', *Journal of Rural Studies*, vol 13, no 3, pp275–288

Johnston, M. and Shimada, L. (2004) 'Urban forestry in a multicultural society', *Journal of Arboriculture*, vol 30, no 3, pp185–196

Kaplan, R. (1973) 'Some psychological benefits of gardening', *Environment and Behavior*, vol 5, pp145–165

Kaplan, R., Austin, M. and Kaplan, S. (2004) 'Open space communities: Resident perceptions, nature benefits and problems with terminology', *Journal of the American Planning Association*, vol 70, no 3, pp300–313

Kaufman, H. (1959) 'Toward an interactional conception of community', *Social Forces*, vol 38, no 1, pp8–17

Kaufman, H. and Wilkinson, K. (1967) *Community Structure and Leadership: An Interactional Perspective in the Study of Community*, Research Bulletin 13, Mississippi State University, Mississippi State, MS

Kellert, S. (1984) 'Urban American perceptions of animals and the natural environment', *Urban Ecology*, vol 8, pp209–228

Kollin, C. (1987) 'Citizen action and the greening of San Francisco', in A. Phillips and D. Gangloff (eds) *Proceedings of the Third National Urban Forestry Conference*, American Forestry Association, Washington, DC

Kornblum, W. (1983) 'Racial and cultural groups on the beach', *Ethnic Groups*, vol 5, pp109–124

Kreger, J. (1973) 'Ecology and black student opinion', *Journal of Environmental Education*, vol 41, no 4, pp30–32

Kretzman, J. and McKnight, J. (1993) *Building Communities from the Inside Out: A Path toward Finding and Mobilizing a Community's Assets*, The Asset-based Community Development Institute, Institute for Policy Research, Northwestern University, Evanston, IL

Kuo, F. (2001) 'Coping with poverty: Impacts of environment and attention in the inner city', *Environment and Behavior*, vol 33, no 1, pp5–34

Kuo, F. (2003) 'The role of arboriculture in a healthy social ecology', *Journal of Arboriculture*, vol 29, no 3, pp148–155

Kuo, F., Bacaicoa, M. and Sullivan, W. (1998) 'Transforming inner-city landscapes: Trees, sense of safety and preferences', *Environment and Behavior*, vol 30, no 1, pp28–59

Lipkis, A. and Lipkis, K. (1990) *The Simple Act of Planting a Tree*, Jeremy Tarcher, Los Angeles, CA

Luloff, A. and Swanson, K. (1995) 'Community agency and disaffection: Lessens for enhancing collective resources', in L. Beaulieu and D. Mulkey (eds) *Investing in People: The Human Capital Needs of Rural America*, Westview Press, Boulder, CO

Maslin, M., Vu, P. and Kidd, C. (1999) *Philadelphia Green Presents the Tree Tenders Handbook*, The Pennsylvania Horticulture Society, Philadelphia, PA

McDonough, M., Burch, W. and Grove, M. (1991) 'The urban resource initiative', in P. Rodbell (ed) *Proceedings of the Fifth National Urban Forest Conference*, American Forestry Association, Washington, DC

McKinney, C. (1985) 'Toxic hazards and blacks', *Black Enterprise*, November, pp23–32

Meeker, J., Woods, W. and Lucas, W. (1973) 'Red, white and black in the national parks', *North American Review*, vol 258, no 3, pp3–7

Mitchell, R. (1980) 'How soft, deep, or left? Present constituencies in the environmental movement for certain world views', *Natural Resources Journal*, vol 20, no 2, pp345–358

Mohai, P. (1990) 'Black environmentalism', *Social Science Quarterly*, vol 71, no 4, pp745–765

Mohai, P. and Bunyan, B. (1999) 'Is there a race effect on concern for environmental quality?', *Public Opinion Quarterly*, vol 62, pp475–505

Mowen, A. and Kerstetter, D. (2006) 'Introductory comments to the special issues of partnerships: Partnership advances and challenges facing the park and recreation profession', *Journal of Park and Recreation Administration*, vol 24, no 1, pp1–6

Nisbet, R. (1953) *The Quest for Community: A Study in the Ethics of Order and Freedom*, Oxford Press, New York, NY

Nowak, D., Noble, M., Sisinni, S. and Dwyer, J. (2001) 'People and trees: Assessing the United States' urban forest resource', *Journal of Forestry*, vol 99, no 3, pp37–42

Park, P. (1992) 'The discovery of participatory research as a new scientific paradigm: Personal and intellectual accounts', *The American Sociologist*, vol 23, no 4, pp29–42

Petras, E. and Porpora, D. (1993) 'Participatory research: Three models and an analysis', *The American Sociologist*, vol 24, no 1, pp28–42

Pettit, K., Kingsley, T. and Coulton, C. (2003) *Neighborhoods and Health: Building Evidence for Local Policy*, The Urban Institute, Washington, DC

Philadelphia Health Monitoring Corporation (2002) *Health Database Household Survey 2002*, Philadelphia, PA

Philipp, S. (2000) 'Race and the pursuit of happiness', *Journal of Leisure Research,* vol 32, pp21–124

Philpott, T. (1978) *The Slum and the Ghetto: Neighborhood Deterioration and Middle Class Reform, Chicago 1980–1930*, Oxford University Press, New York, NY

Piehl, A. and Zhang, J. (2003) *Lancaster Avenue Urban Design Study*, Hamer Center for Community Design, University Park, PA

Reardon, D. (1993) 'Participatory action research from the inside: Community development in East St Louis', *The American Sociologist,* vol 24, no 1, pp69–106

Rios, M. (2002) 'Building social capital through participatory design', *New Village*, no 3, pp18–26

Rios, M., Gabler, M. and Marquardt, S. (2002) *West Philadelphia: Sarah Allen Community Workshop*, Penn State Hammer Center for Community Design Assistance, University Park, PA

Rudel, T. (1989) *Situations and Strategies in Land Use Planning*, Cambridge University Press, New York, NY

Salazar, D. (1996) 'Environmental justice and a peoples' forestry', *Journal of Forestry*, vol 94, no 11, pp32–36

Sanoff, H. (2000) *Community Participation Methods in Design and Planning*, John Wily and Sons, New York, NY

Schelhas, J. (2002) 'Race, ethnicity and natural resources in the United States: A review', *Natural Resources Journal*, vol 42, no 4, pp723–763

Schinew, K., Floyd, M. and Parry, D. (2004) 'Understanding the relationship between race and leisure activity and constraints: Exploring an alternative framework', *Leisure Science*, vol 26, no 2, pp181–199

Shabecoff, P. (1990) 'Environmental groups are told they are racists in hiring', *The New York Times*, 1 February, p11

Sirianni, C. and Friedland, L. (2001) *Civic Innovation in America: Community Empowerment, Public Policy and the Movement for Civic Renewal*, University of California Press, Berkeley, CA

Sheppard, J. (1995) 'The black–white environmental gap: An examination of environmental paradigms', *Journal of Environmental Education*, vol 26, no 2, pp24–35

Shogan, R. and Thomas, C. (1964) *The Detroit Race Riot*, Chilton, New York, NY

Shutkin, W. (2000) *The Land That Could Be: Environmentalism and Democracy in the Twenty-First Century*, The MIT Press, Cambridge, MA

Taylor, A., Wiley, A., Kuo, F. and Sullivan, W. (1998) 'Growing up in the inner city: Green spaces as places to grow', *Environment and Behavior*, vol 30, no 1, pp3– 27

Taylor, D. (1982) *Contemporary Opinions Concerning Environmental Issues*, Northeastern Illinois University, Chicago, IL

Taylor, D. (1989) 'Blacks and the Environment: Toward an explanation of the concern and action gap between blacks and whites', *Environment and Behavior*, vol 21, no 1, pp175–205

Ulrich, R. (1984) 'View through a window may influence recovery after surgery', *Science*, vol 224, pp420–421

US EPA (United States Environmental Protection Agency) (2002) *Community, Culture and the Environment: A Guide to Understand Sense of Place*, EPA 842-B-01-03, US EPA, Washington, DC

Van Liere, K. and Dunlap, R. (1980) 'The social bases of environmental concern: A review of hypothesis, explanations and empirical evidence', *Public Opinion Quarterly*, vol 44, summer, pp181–197

Virden, R. and Walker, J. (1999) 'Ethnic/racial and gender variations among meanings given to and preferences for the natural environment', *Leisure Science*, vol 21, pp219–239

Wallerstein, D. (1997) 'Freirian praxis in health education and community organizing', in M. Minker (ed) *Community Organizing and Community Building for Health*, Rutgers University Press, New Brunswick, NJ

Walter, A. and Crenshaw, R. (1992) *Sustainable Cities: Concepts and Strategies for Eco-City Development*, Eco-Home Media, Los Angeles, CA

Warren, R. (1972) *The Community in America*, Rand McNally, Chicago, IL

Washburne, R. (1978) 'Black participation in wildland recreation: Alternative explanations', *Leisure Sciences*, vol 2, pp175–189

Whyte, T. (1989) 'Participatory action research: Through practice to science in social research', *American Behavioral Scientist*, vol 32, no 5, pp513–551

Wilkinson, K. (1979) 'Social well-being and community', *Journal of the Community Development Society*, vol 10, no 1, pp5–16

Wilkinson, K. (1991) *The Community in Rural America*, Greenwood Press, New York, NY

Wilkinson, K. (1998) 'The community: Its structure and process', in E. Zuber, S. Nelsen and A. E. Luloff (eds) *Community: A Biography in Honor of the Life and Work of Ken Wilkinson*, The Northeast Regional Center for Rural Development, University Park, PA

Wolf, K. (1998) *Human Dimensions of the Urban Forest*, Center for Urban Horticulture, University of Washington, Seattle, WA

5

Creating Common Ground: A Collaborative Approach to Environmental Reclamation and Cultural Preservation

Jacquelyn Ross, Shannon Brawley, Jan Lowrey and Don L. Hankins

This is the story of how an undergraduate's small landscape plan grew into a complex reclamation project in the form of a riparian/wetland-based garden in Yolo County, California. From the beginning, the work was a collaborative process. The use of participatory research opened communication and created common ground between competing interests. The project grew organically out of a student's desire to be inclusive of local community stakeholders. The Native American Tending and Gathering Garden (the Garden) is in the Cache Creek Nature Preserve (CCNP) located in Woodland, California, which is managed by the Cache Creek Conservancy (the Conservancy) (see Figure 5.1). The Garden is the result of collaboration between industry, the Native American community, academics, farmers and others. The journey that led to the establishment of the Garden was laden with lessons. This chapter provides a critique of the impact and contribution of participatory action research (PAR) in a local community-based natural resource management effort.

HISTORY: SETTING THE CONTEXT

The genesis of the Conservancy itself was steeped in decades' long local controversy surrounding gravel mining on Cache Creek. The resolution of that controversy indirectly created the collaborative atmosphere that welcomed the Garden concept. A review of the Conservancy's history, as well as Native American participation

Source: courtesy of Green Info Network

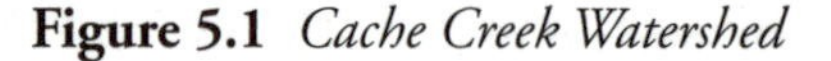
Figure 5.1 *Cache Creek Watershed*

within the watershed, illustrates why the Garden concept (and all that resulted) was able to find a home at the CCNP.

The Cache Creek watershed drains 420 square miles (1088 square kilometres) of the Coast Range as it winds eastward along a 100 mile (161km) course through California's Lake and Colusa counties, then southeast through the Capay Valley before its confluence with the Sacramento River. The creek is recognized on the National Register of Historic Places for its cultural richness and importance. Historically, several different groups of indigenous tribal peoples made the Cache Creek watershed their home, including Miwok, Patwin, Pomo, Wappo and Wintun peoples. Native Americans had a significant presence in the Cache Creek watershed before and during initial European settlement. Although traces of native village sites dot the banks of the Cache Creek, the Rumsey Band of Wintun Indians is the one remaining tribe with a land base in the watershed today.

Native American community members continue to live in the area; but there has been little recognition of local tribal expertise or knowledge of natural resource management and history. The county has grappled with natural resource management issues since the early 1950s. There is no evidence that those who knew the creek's pre-contact condition were asked for their input or help. During the early 1980s, the Rumsey Band of Wintun tribal members sought to re-establish a tribal land base in the Capay Valley and bring the Wintun families home from the places to which they had migrated. Local news articles chronicled the apprehension and attitudes of some non-native residents in the valley. In Woodland, the seat of Yolo County, the city newspaper ran an article entitled 'Indians' return stirs Capay Valley protest'. A retired local medical doctor was quoted: 'I don't like the idea of having drunken Indians up and down the highway... The Indians will steal anything around' (Dianda, 1981). While such sentiments are not universal, countervailing public opinion has been largely absent. Community dialogue about the changing use of the Cache Creek is extensive; yet there has been scant acknowledgement of the impact upon Native life ways in either the historic or contemporary context.

Private landownership along the creek and loss of riparian landscape greatly diminished land access for native peoples. Access is necessary for traditional food gathering, hunting and ceremony. It is required for the tending and harvesting of plants necessary for the creation of baskets, traps, cordage and other uses. Under similar conditions in other parts of the US, lack of access has affected Native American people in a variety of ways (Anderson, 2005; Turner, 2005). In the context of Cache Creek, there has been no such discussion and, hence, no protection for health and cultural concerns unique to tribal peoples of the area.

European settlement has dramatically impacted upon the landscape. Among the local impacts was the diversion of water from the creek in 1856 into a canal that was the predecessor of the current system of dams and canals, which diverts the creek into a countywide water delivery system for agriculture and other uses within Yolo County. Upstream of the agricultural diversions the creek remains much as it has always been, albeit with the significant invasion of exotic plant species. Downstream of the diversions, the creek became a major source of aggregate (sand and gravel) starting in the late 1930s and intensifying during the 1970s and 1980s. The gravel industry began on Cache Creek as small family-run operations. One of those small companies is now owned by Rinker Materials, the second largest construction materials company in the world. Each gravel mining operation has grown rapidly, working to meet the demand for new construction in California. The Bay Bridge in the San Francisco Bay Area contains Cache Creek materials. Mining within the creek's active channel grew from a few hundred thousand tons (1 ton is roughly equivalent to 1 cubic yard) during the 1950s to 5 million tons annually under a new permitting process established in 1997.

The riparian corridor changed. Some riparian landowners were angered because they were losing their land to erosion. Environmentalists joined the fray. Disagreements turned into feuds that lasted decades. From 1974 to 1997, two decades of 'gravel wars' accomplished virtually nothing for Yolo County. In 1994, newly elected county leadership marshaled the courage and vision to face the state-mandated challenge of devising a plan to allow continued mining while fostering reclamation and restoration. The Cache Creek Area Plan, which included the Cache Creek Resources Management Plan (CCRMP), was conceived in the context of 'net gain'. This concept is based on the idea that the people of Yolo County and their natural resources would be better off at the end of 30 years of mining permits than if there had been no mining at all. In order to accomplish the net gain goal, the county would permit mining, industry would provide remediation funding, and a community collaboration would be formed. A plan for managing Cache Creek's resources to the benefit of all would result from this work.

Through the process that created the CCRMP, a sense of community purpose was kindled and disparate interests started identifying common goals. Farmers helped miners to restore off-channel pits to more fertile conditions than before mining. Gravel mining companies were recognized for the flood control and erosion control work they had provided for decades. Additionally, the companies agreed to contribute 20 cents for each ton of gravel mined for creek restoration. At 5 million tons mined per year, their yearly contribution toward environmental improvement would total US$1 million. Starting in 2008, the gravel industry will be increasing to 45 cents per ton. The beneficiary of these proceeds is the Conservancy.

THE CACHE CREEK CONSERVANCY

The Conservancy was created as a vehicle for implementing the CCRMP. The initial board of directors included gravel miners, local government representatives, farmers, small business owners, university professors and a local historian. The Conservancy was the first organization in living memory dedicated to restoring Cache Creek's riparian corridor in the area historically mined for gravel. Its mission is to promote the restoration, enhancement and prudent management of the stream environment along Cache Creek from Capay Dam to the Settling Basin just east of Woodland. Created by the Army Corp of Engineers, the basin is a sink that lets sediment drop before the creek reaches the weir and levee that control water before it goes into the Yolo Bypass and then on to the Sacramento River:

The Conservancy's mission statement was formed in the following context:

> *the citizens of Yolo County acknowledge that Cache Creek is a valued resource. Past activities, including agriculture, mining, groundwater extraction, damming, irrigation discharge, other infrastructure development and construction along the creek have modified its wildlife habitat values. With the Cache Creek Resource Management Plan approved by the board of supervisors, there is now an opportunity for a coordinated response to revitalize the riparian habitat along Cache Creek. The Cache Creek Conservancy has been created to be a focal point for accomplishing many of the habitat projects identified in the management plan. (Cache Creek Conservancy Board of Directors, 1999)*

The Conservancy searched for a restoration project and found that building trust and creating relationships with landowners were necessary first steps. No private landowner offered to allow a restoration project on their property by an organization that was untried, untested and distrusted. Local landowners could not decide whether the Conservancy was a bunch of environmentalists, a trick by the government to find infractions of mining regulations and inflict penalties, or a sham by gravel miners who had pulled the wool over everyone's eyes to get their use permits renewed.

The new Conservancy board members bolstered project efforts with personal calls to longtime friends who were also riparian landowners, building trust and opening the door for projects. Board members placed their reputations on the line in the aftermath of 20 years of mining discord within the rural community. In 1997, the first private landowner offered her land for a restoration project. It was a small area – nearly invisible; but this project was successfully completed a year later. The surrounding community watched how the landowner was treated by the Conservancy, looking to see if she and her land were respected and whether the result was worth the effort. The Native American community would assess the Conservancy in much the same manner some three years later.

In 1999, one of the gravel companies offered to donate a 130 acre property to the county if the Conservancy would assume management. The next 18 months of negotiations included site planning, a conservation easement to be held by the Conservancy and legal agreements. What evolved from these negotiations was the Cache Creek Nature Preserve, now the Jan T. Lowrey Cache Creek Nature Preserve, created in a context of community collaboration and involvement. The success of the Conservancy and the CCNP indicated a predisposition for community members' involvement in a collaborative process. The community networking, a partnership of the Conservancy and the CCRMP, played a crucial background role in the evolution of the Garden. Since the Conservancy was born in an atmosphere of controversy, board members and staff learned that building trust through carefully designed and implemented projects was the key to success.

In the beginning, University of California, Davis, student Shannon Brawley's idea for a garden of culturally important local plant species was an academically controversial project. There were detractors within the academic community who questioned whether a project such as this one held validity. On the community end, the collaboration of the wary Native American community in the local area would need to be an essential component for the project to succeed and have value.

In this case, one must ask if adversity sets the groundwork for creativity and collaboration. Had the 'gravel wars' never taken place there would have been no CCRMP. Without the Resources Management Plan there would have been no Conservancy. The Conservancy's successes with the local landowners led to the donation of property, which became the CCNP. Without this preserve, there would have been no place for the concept of a garden to take root. Taken in a historical context, the Garden is the result of a logical collaborative process. The success of Brawley's project is partially the result of all that came before it and partially the result of a good idea embraced by dedicated communities who had grown weary of discord. The context of a project is shaped by the participants and is influenced by their attitudes. This may be a fundamental deciding factor in the success of any given participatory research effort.

The genesis of the garden

In 2000, Brawley conceptualized the Garden as she finished her undergraduate program in landscape architecture and continued in the geography PhD program. Having read of the extensive environmental management utilized by Native Americans in California, the restoration impact of such practices intrigued her. In particular, she took note of the management skills of California Native basket weavers. Furthermore, she learned that today's weavers have difficulty in accessing traditional basketry plants, such as willow (*Salix* spp) and deergrass (*Muhlenbergia rigens*) that are free from pesticides and other chemical contaminants. Brawley contemplated two questions: could something come from a scholarly investigation of this problem? Would weavers be interested in participating in the research? From the literature (see Anderson, 2005) and later through conversations with cultural practitioners, she learned that traditional management occurred at various spatial and temporal scales (e.g. from the individual to the ecosystem and from multiple times within a season to years between management actions).

Native land managers used a number of different traditional management techniques: coppicing, pruning, tilling, transplanting, weeding and prescribed burning. These management practices mimic natural disturbances, such as lightning-induced fires, floods and animal activity. Each technique differs depending upon the scale needed and the seasonal application. Traditional management helped to maintain the plant and animal populations essential to native peoples' way of life, while supporting habitat diversity. Traditional native management tools can be the foundation for modern management practices, which will help land managers to conserve habitat biodiversity.

The native concept of land stewardship sparked Brawley's idea of creating a garden focused on a core of local native plants. Redbud (*Cercis occidentalis*), willow (*Salix* spp), tule (*Schoenoplectus acutus*) and other plants would be tended by traditional native management techniques, such as burning, coppicing, pruning and thinning. Brawley believed that she could learn from the weavers and other Native American land managers; but there was scant precedent for their active participation in the research process.

In formulating her plan to work with the native community, Brawley sought the advice of her university teachers, but found little enthusiasm for working with native people. One of her professors told her 'people want to move into the 21st century'. This same faculty member urged a tight focus on pure scientific environmental research based on quantifiable data and statistical evidence. He discouraged the proposal to integrate applied culturally based environmental knowledge. He did not recognize the suitability of both quantitative and qualitative data collection for this project. One advantage of utilizing PAR is that both ways of accomplishing data collection are accepted. Qualitative data is especially important to promoting action because it considers the experience of individuals in the community. Recounting stories and experiences can be galvanizing. This is why the project steering committee utilized various approaches to gathering information.

Other faculty members were more encouraging to Brawley. One teacher gave her the phone number of the California Indian Basketweavers Association (CIBA). This group is the first arts service organization of its kind for Native American weavers in the US. CIBA staff connected Brawley with master weaving teacher Kathy Wallace, a descendent of the Karuk and Yurok peoples, and a member of the Hoopa Tribe, all of which are native nations based in northern California. Another of Brawley's teachers introduced her to the CCNP, located in Yolo County, California, as a possible home for the garden. The 130 acre preserve includes a 28 acre wetland, a reclaimed aggregate mining pit, oak savannah and a section of riparian corridor. The Conservancy's Executive Director Jan Lowrey greeted the idea with enthusiasm. He suggested a 2 acre site by the wetlands in the preserve as an optimal site for Brawley's project. The garden would serve as an important addition to the CCNP educational program, addressing the continuing presence and practices of Native American communities within the Cache Creek watershed. Unlike any other education program in the local area, an ethno-botanical basketry garden represented a cross-cultural approach to hands-on environmental and cultural education emphasizing the relationship between plants and people. Brawley presented the idea to the Conservancy's board of directors. There was some apprehension, but the majority of the board seemed to embrace the idea. She received authorization to proceed with her plans.

Wallace would prove to be a pivotal contact. She visited the potential 2 acre garden site off the wetlands area that had once been a gravel mining pit. She helped to refine design ideas for the Garden. She encouraged a visit to the local tribe to introduce the project idea as this was an important first step in terms of traditional protocol and respect. An initial meeting between Brawley, Rumsey Band of Wintun tribal chairwoman Paula Lorenzo and three women weavers from the region set the stage for a project visioning process, which began in 2000. These women greeted the garden concept with a mixture of hesitancy and support. An elder Wintun master weaver offered excellent ideas about the garden design. For example, she wanted to see a shade structure built so weavers could retreat out of the sun when processing materials. She also suggested implementation of a special garden for children. She felt that children should develop respect and knowledge about the environment before moving on to weaving. Such ideas helped to complete the conceptualization of the garden and increase its cultural utility.

Wallace's experience and optimism continued to motivate Brawley and together they formulated a guest list of native weavers and cultural teachers for an open house/community forum to discuss the Garden project. Most of the people who came to the forum had positive experiences in working with each other on other projects. New suggestions for the Garden were offered quickly, such as the addition of a fire pit and curriculum for educating visitors. The note takers at the forum were hard pressed to keep track of all the ideas. Attendees added their choices to an ever expanding wish list of plants that was passed from person to person. They debated about the feasibility and desirability of additional project elements such as a living willow fence to distinguish between separate gathering areas for youth and adults. At the end of this forum, a list of ten priorities, ranging from the implementation of the garden, management of the garden, outreach curriculum, docent training and internships, was adopted for the project. As a group, the guests quickly asserted themselves as designers, planners and policy-makers for the venture and became a governing body now known as the Tending and Gathering Garden (TGG) Steering Committee (the Committee). No one at the CCNP anticipated this high degree of participation from these Native cultural practitioners. Clear, forthright, highly welcoming and inclusive communication was the bridge necessary to engage a community traditionally absent from academic discourse.

The Rumsey Community Fund (the philanthropic arm of the Rumsey Band of Wintun Indians) and the Teichert Foundation, a local gravel industry non-profit organization, donated money to implement the Garden. Some of these initial funds supported the open house/community visioning forum that allowed the Conservancy staff, weavers and cultural practitioners to discuss the project together. Currently, the Rumsey Community Fund has funded the majority of the Garden's implementation, as well as a project coordinator position. Brawley was asked by the Committee to be the coordinator until the project implementation was complete.

THE GARDEN PARTICIPATORY ACTION RESEARCH (PAR) APPROACH

This project did not start out as an intentional PAR endeavor. The participatory process was well under way when Brawley first introduced the term 'participatory action research' to the Committee. Yet, the work at the Garden is participatory research of high order. Brawley worked closely with a few Committee members, discussing the research problem, questions, methodology and data analysis. They then brought their concepts to the full committee for discussion and consensus. All were concerned that the research results would have practical utility and meet the needs of the native community.

The Committee members hail from 14 different tribes, including California tribes such as Maidu, Yurok and Pomo. In addition to their many skills and broad knowledge base as cultural practitioners, committee members brought a variety of professional skills from their positions as biologists, weavers, artists, policy designers, teachers, writers and account managers. Others who worked on the committee were Conservancy staff members Jan Lowrey, the Garden project coordinator/researcher Brawley and the CCNP education coordinator. This diversity added dimension to discussion and problem-solving, providing the needed advice to make this project a success. It is noteworthy that the majority of the committee was and is composed of women. This is not the result of exclusion, but rather a reflection of the proportion of female to male native weavers in California today. Drastic changes in hunting and fishing access and the introduction of government regulation of these activities have had a deleterious effect on traditional life. The weaving of utilitarian baskets, nets and traps was once the specialty of the men.

A participatory research process needs to be flexible to accommodate the schedules and other activities of community participants. For the Garden, these needs are accommodated in several ways. The Committee meets in the evenings and on weekends so that minimal time is lost from full-time jobs. Committee members are volunteers and their limited time is respected by flexible scheduling outside of regular business hours. The meetings often include potluck meals as some members travel far to get to the evening and weekend meetings. Members who miss meetings or events receive reports on all proceedings. Although agendas and minutes are a part of each meeting, the Committee does not use parliamentary procedure as the decision-making process. From the beginning, the committee employed a discussion and consensus process, and defers to elders. Traditional native governance often values peacekeeping, good community relations and long-range planning. The Committee meetings are always respectful, enjoyable and often several hours long. The relaxed, convivial atmosphere creates a space where everyone is allowed time to say what they need to share. Participatory research welcomes community ethics, culture and worldviews. These components are in regular practice in this project and have fostered respect and understanding within the collaborative. It is common for the Committee members, CCNP staff and project volunteers to enjoy lengthy, informal discussions following meetings and events. These discussions are important because they increase trust and cross-cultural sharing.

Projects involving native communities seem to appeal to a wide range of people. Sometimes this interest goes awry. The Garden's native partners have experience with outsiders trying to appropriate tribal knowledge through research or grant-seeking ventures. Commonly, there was no reciprocal contribution or credit given to the host community. The sharing of these negative experiences with the non-native members of the Committee helped them to understand how mistakes had been made with tribal communities. Non-native committee members initially approached the project with a certain naiveté about project ownership, believing perhaps that issues of academic acknowledgment and intellectual property ownership would not be important issues for the native community.

Anyone contemplating cross-cultural work must be vigilant in respecting, honoring and protecting the traditional knowledge held by individuals and communities. There are boundaries to be recognized so as not to allow the research to become just another extractive process. It is imperative to build relationships that foster a mutual quest for knowledge and understanding (Simpson, 1999, 2000; Wilmsen, forthcoming). For example, Brawley made a conscious decision neither to interview nor beseech a local Wintun elder to collaborate in this project even though the elder has vast basket weaving and environmental knowledge. When the elder did choose to participate, she defined her own boundaries and let it be known that she would leave the project if things were not done correctly. In the past, she had shared her knowledge generously with outsiders, not knowing that they would later publish this information. This was tremendously painful to her. Scholars cannot assume that a community has familiarity with or innocence about standard research practice. The research and the Garden project benefited from disclosure about publication and research protocol because all involved have a responsibility to the communities they represent. Understanding Brawley's academic responsibilities and meeting her faculty advisers helped Committee members to support her work because they had a fuller picture of university expectations. Revisiting these issues has helped to keep the focus on both the community utility and academic requirements of the project.

The research problem/need

California's rural communities, as well as rural communities around the world, are facing a collision of interests and needs that often leave indigenous people out of the planning process. Policy-makers have to invest in long-term strategies that will facilitate conservation of the environment while considering sustainable economic opportunities for local rural populations.

Throughout the state of California, 'access to basketry materials has been limited by private property boundaries, as well as by public land laws and management practices which preclude gathering' (Ortiz, 1993). For instance, California's Central Valley wetlands and riparian ecosystems have been reduced by 90 per cent since 1850, due in large part to human impacts (Barbour et al, 1993). This, to some extent, is due to population growth, the encroachment of housing, business and agricultural development in the region. Riparian communities along Cache Creek are now plagued with invasive plants such as salt cedar (*Tamarix parviflora)* and arundo (*Arundo donax*) that exacerbate flooding and erosion. Interestingly, years ago both species were introduced to control the very issues they seem to cause today.

This set of problems influenced one of the main project objectives for the Garden: the creation of a safe place for native educators and cultural practitioners to teach traditional plant management and gathering techniques. The lack of access to gathering areas makes it difficult to pass on cultural traditions such as basketry to family and community members. Baskets made by California native weavers are assessed as some of the finest anywhere in the world. Exceptional weavers once tended and harvested materials in the Cache Creek watershed. Very little weaving continues in this specific area today, although it is flourishing in several other tribal regions. Historians and ecologists seem to have missed the intricacy and refinement of the relationship between tribal people and the land. Such omissions have impacted upon the land and, thus, the native peoples throughout the state. Indigenous land management techniques leave subtle marks. Only now are Western scientists scratching the surface of this knowledge.

In recent years, due in large part to the efforts of CIBA, some state and governmental agencies have granted permission to collect basketry resources, such as beargrass (*Xerophyllum tenax*), on public lands. CIBA has also focused attention on the toxic effects of herbicides and pesticides on weavers' health. Weavers often hold plant materials in their mouths to aid in splitting the fibers into proper strands. Basketry plants are highly hand processed, so both topical and systemic application of chemicals in natural environments are of concern. The continuation of weaving traditions is important to contemporary weavers and requires plentiful high-quality plant sources. Although the Garden's Committee members have access to their personal gathering areas, they are having a difficult time locating additional natural areas where potential weavers can be taken to identify plants and learn traditional management techniques.

Research objectives and process

The research questions evolved from Brawley's initial set of questions. She and a Committee member grappled with the direction of the research and formulated new questions that focused on the reclamation of the garden site and its ultimate utility to educators, cultural practitioners and the general public. The Committee considered these concerns and refined them to produce the following primary research objectives:

- Document and analyze the process between the Committee and the Conservancy in implementing the Garden reclamation project.
- Document the creation of a safe place for native educators and cultural practitioners to teach traditional plant and management techniques.
- Determine the optimal methods for environmental mitigation for those collaborating on community-based restoration and land management projects similar to the Garden.
- Study the effects of fire on Santa Barbara sedge (*Carex barbarae*) and changes in overall plant densities before and after application of fire.

As the project began to unfold, so did the research. One difficulty faced in the beginning was defining and understanding what PAR is. Additional reading and training sessions such as the Community Forestry and Environmental Research Partnerships (CFERP) annual research workshop helped project members to define our research. CFERP included community members in the workshops, which allowed the community members an opportunity to educate students and faculty, and to share experiences with each other. Workshops like these helped to bolster the researcher and the community members when they went back home to continue the work.

Methods and analysis

Both quantitative and qualitative methods of data collection were used to respond to the research questions. The first objective was to document and analyze the process between the Committee and the Conservancy in implementing the Garden reclamation project. Documentation of this process was achieved through tape and video recordings, design charettes and written documents, such as meeting minutes, mapping of stakeholder relationships and grants, and billing information for the length of the project's research implementation (2000 to early 2005).

One of the most effective ways of gathering information and one that seemed to come naturally for the Committee was using a whiteboard to map out ideas or thoughts visually. For example, one evening three members met to discuss a writing project. Jan Lowrey, the Conservancy's executive director, was also a fifth-generation local farmer. He started to map a history of the watershed and the relationship between local Native Americans and non-native peoples. What the three Committee members began to notice was that the map became two separate but parallel sections that demonstrated how these two groups of people lived together; but the Native American presence was invisible and ignored until the modern residential housing for the Wintun became an issue. During this process, members drew upon Jan's memories of growing up in the area and old archival clippings he had in the office. The conversation evolved into a mapping of trust and risk issues for all of the stakeholders involved in the Garden. This became a pivotal document for reframing conversations.

Communication in this project occurs on several levels, partly because the project is cross-cultural and partly because of the hierarchy and protocol within the Conservancy and the Conservancy's relationship with stakeholders. On one level, the common interaction between the Committee and the Conservancy continues to be conducted via email, phone and mail. Most individuals on the Committee have email and those that do not are kept informed by mail or by simply picking up the phone and calling. Email provides a quick and easy way of reviewing and approving signage, publication and meeting notes. The tape recordings and videotapes of meetings have been a way for the researcher to assess the process. What clearly presents itself in the transcriptions is the way in which the meetings are conducted. Each person is given the opportunity to weigh in on any issue, which gradually leads to a consensus of what direction should be taken. This also leads to creative solutions. The discussion about design of the shade structure within the Garden is a good example of this. The Committee deliberated over whether it should be of traditional or modern design, who would make it, how to insure it and how the public could use the structure. Ultimately, the innovative structure that was created was reminiscent of a traditional ceremonial structure, but adhered to state and county building codes.

The most challenging line of communication has been between the Conservancy board of directors and the Committee. The executive director (Lowrey) and the project coordinator/researcher (Brawley) went to each board meeting and presented quarterly activity reports. The Committee as a whole, however, expressed interest in having a native community member in a position to communicate directly with the board. This was unconventional in a typically hierarchical leadership structure. Eventually, a Committee member, Don Hankins (Miwko and Osage), was appointed to attend board meetings to convey project developments and to report back to the Committee. He was in a position to see issues differently than staff members did and with the added benefit of having a specific cultural lens, as well as professional background as a biologist. For Hankins, it meant a larger time investment in the Garden project. The Committee wanted the board to understand the expertise and the essential contributions of the group. Hankins conveyed that ably.

The Conservancy's support of the Committee was considered unique. No land-managing organization in the area had developed this kind of relationship with Native Americans. Outside groups wanted to know how to replicate this. An important part of the relationship was the Committee's development of specific policies that established structure for the Garden and the working relationship with Conservancy staff and board. These policies helped to define the operation and status of the Garden within the Conservancy. Lowrey defined the need for a respectful relationship in the following manner:

> *For their part, the Steering Committee members have asked for nearly nothing from the project. Many have freely given of their time and expertise for presentations to many diverse audiences, ranging from Cal State Sacramento to the California Mining Association. However, there is a long and disturbing history of individuals, organizations and agencies drawing on the native community's knowledge, regalia and materials without giving credit where credit is due. Therefore, the Steering Committee members ask three things: respect for themselves as the professionals [whom] they are; recognition for their culture's unique contribution to the Tending and Gathering Garden project overall; and control over their intellectual property (that is, the oral history and teachings from generations of elders that cannot be found elsewhere). It is in this context that the Steering Committee members have composed policies and guidelines for the Tending and Gathering Garden.*

The policies are meant to set forth guidelines about the relationship of individual members to the Committee, the Committee to the board of governors, and Garden visitors to the Garden. These policies are currently in final draft, ready to be presented to the board.

The Garden collaborative is multidimensional. It is easy to think of the Committee as a pan-Indian group representing California native people. This is a misconception. The Committee members all come from different tribal nations, with separate ancestries, governments and cultural legacies. Designing rules to govern the Garden represents a dimension of international negotiation. Traditional rules form the policy base. One of the most important rules is the honoring and recognition of the tribal people indigenous to the Cache Creek watershed. As the traditional local stewards of the land, their concerns have priority. The Committee often consults with local cultural practitioners and tribal leaders to make decisions and recommendations about the Garden. Adding complexity, the location of the project on public land means that the culturally foreign political and legal boundaries of local and state regulation must be considered. This diversity of cultures and governing bodies is an important factor in the participatory work. Communication between all parties must be clear and consistent. This takes more time than a conventional research project would. It also requires a bigger investment in relationships. Understanding what motivates each party and participant is important because the Garden is meant to be a permanent feature in the CCNP. It was established carefully so that it can live on through changes in Committee membership, changes in the CCNP and changes in the Conservancy board and staff.

The discussion of specific perspectives has come up often over the past few years. When Conservancy staff, Committee members and the researcher present the Garden project to broader audiences, it is typical that three perspectives are presented. Each presenter describes distinct roles in the research and project. Conservancy staff speak on mining issues and the work of restoring mining sites, the mechanisms within local government that made such projects possible, and the benefits of having multi-tribal perspectives and help in the project. Committee members discuss the tribal history of the Cache Creek watershed, traditional and contemporary needs and usage of plant species in the area, and the importance of being recognized and included in the restoration work. Brawley speaks about what she learned in applying her academic skills to a real world problem and the needs of a community. She also addresses the results of the specific plant restoration efforts and the Garden design. At gravel and mining industry events, it is not uncommon to see industry leaders respond to this presentation and publicly support this kind of creative environmental collaboration.

The second project objective planned by the Committee was to document the creation of a safe place for native educators and cultural practitioners to teach traditional plant and management techniques. It was determined that by combining mechanical methods (including harrowing and irrigation) and traditional management techniques, a site can be reclaimed to a high level of cultural utility. The Committee asked Brawley to document site analysis, plant choices, planting design, construction plans, weed management, soil preparation, irrigation and cost analysis. Photo site recording was also conducted during the implementation of the Garden.

The Garden was divided into four manageable pieces, which made it easier to plant and manage for weeds. A bed of sedge (*Carex barbarae*) was the first section to be installed. Hankins, Brawley and Lowrey analyzed the existing soil and immediately knew that the soil would not produce the long white roots necessary for basketry due to the high quantity of gravelly soils present. The Garden soil would need to be amended with sandy loam soil. With this soil improvement, the Garden was ready for sedge plants. In June 2001, over three consecutive days, 2000 sedge plugs were planted by schoolchildren from Yoche-de-he School (the tribal school of the Rumsey Band of Wintun Indians), the California Conservation Corps and Conservancy staff members. Planting anything in June was a huge risk due to the possibility that the plants would die in the summer heat. Plants received water from a fire hose and soaker hose every day throughout the summer until the sedge was established. Initially, it appeared that around the periphery of the sedge bed, 89 plants had died. They reappeared with the spring rains. Thus, the first lesson was to plant with the fall and winter rains. In the fall of 2002, juncus (*Juncus* spp), dogbane (*Apocynum cannabinum*), blue wild rye (*Elymus glaucus*), creeping wild rye (*Leymus triticoides*) and another variety of sedge were introduced to the Garden. All of the blue wild rye fell to a voracious flock of Canada geese.

The geese seem to be particularly attracted to blue wild rye and barley that were planted over the years.

Weed eradication was a priority for the reclamation to thrive as a native habitat. No herbicide was used on the 2 acres at the request of the Committee. With the help of local farmers, Brawley established a process of watering the flat area of the site utilizing local farmers' field irrigation pipes. The deep watering that this type of irrigation provided helped the Garden plantings enormously. With every new flush of weeds, local farmers would harrow the soil with their tractors. This project benefited greatly from such generosity. The tree plantings, as well as the grasses, flourished and were able to set root in the gravelly soil.

At the time of writing, the garden implementation continues to bring together experts such as Committee members, farmers, Conservancy staff and industry restorationists who offer advice on plant placement, soil preparation and irrigation. The local farmers continue to provide the use of their farm equipment for irrigation and the harrowing of weeds. The use of their expensive farm machinery has made weed management in the garden much easier. Teichert's restorationists supply plants from the watershed for the Garden and they have shared their planting techniques in gravel overburdened soils. Their field experience contributes to the garden's success. This also represents local community reciprocity.

Among the traditional management techniques implemented in the Garden, the use of prescribed burning attracted special attention. Although this was a common pre-contact tool for managing the landscape in California, contemporary fire suppression policy in the state meant that many local people had never seen a burn for restoration purposes. Brawley and Hankins presented the Committee with a study idea to ascertain the effects of fire on Santa Barbara sedge. They wanted to see if there would be any changes to the plant density before and after a wet season (winter) fire was applied.

The experiment was mapped out on the whiteboard and the whole Committee discussed the idea. All the members and visiting guests contributed. The main concern was that weavers would need an area that was left unburned for gathering while the study was conducted. A visiting guest questioned why sedge should be burned. Collectively, the committee felt that sedge is a plant within the riparian ecosystem that would have been burned when this management technique was utilized traditionally within the region. The committee also explained the cultural significance of sedge as a basketry plant for many tribal groups and their weavers. Most importantly, a study like this one had never been done before.

The Committee also discussed extending the utility of the research. This information would be useful to agencies in which large sedge populations were located (currently, the National Park Service's Pinnacles National Monument is interested in the Garden results). The Committee supported the research idea and study. The preliminary findings indicate that the overall plant success and cultural utility of the garden have been achieved.

Long-term data collection

Included in the general Garden policy is a requirement for gatherers to contribute to data on amounts of plant material harvested, field observations about plant health and environment, and harvesting techniques. This quantifiable data will provide hard evidence about management technique effects in the Garden ecosystem. Qualitatively, Garden user surveys will provide data about visiting weavers and their observations. Review of this material will assist the Committee in evaluating the health of the Garden. It will also be useful information for evaluating the project's utility to the native community and establishing outreach targets for the education component. For example, since the implementation of a data collection process, the Committee knows that the majority of weavers who have gathered sedge have used small garden tools (e.g. trowels) to harvest. Most weavers have felt the quality of roots they gather in the garden is good with respect to length, color and ease of harvesting. One weaver commented: 'Looking forward to coming back and finding the six footers (desirable, long roots) – the place looks good!' Data notes from weavers who are using the Garden reveal that the soil needs to be augmented with more sandy soil and organic matter. The majority of individuals who have been gathering in the garden are elders. Their approval and expert observations are invaluable.

Presenting and sharing the research

Outreach remains an important component of Garden operations. Many visitors come to the CCNP for interpretive events and outdoor education. As the Garden is located prominently in the core section of the preserve, an interpretive program and curriculum are important to explain this project thoroughly and to ensure that the area is treated with respect. The Committee has produced an interpretive brochure and multilingual signage (Wintun, English and Latin) in the Garden. Curriculum modules are to be finalized over the coming months. Considerable effort goes into describing the significance of the collaborative research to the Native American community, industry and the academy. This helps demonstrate to visitors and audiences that, even with disparity, multiple stakeholders can come together over issues such as restoration, education and cultural preservation.

COMMUNITY ACTION IN COMMUNITY-BASED NATURAL RESOURCE MANAGEMENT

Through participation in the Garden and related research, contemporary history has begun to turn and Native Americans are now contributing their traditional knowledge and voices to the management of the Cache Creek watershed.

Participatory research is often said to produce community ownership of the research (Reason, 2006; Pain, 2007). Thus far, this is the case with the Garden. Community members are forthcoming in offering their individual strengths. They have assumed an enormous amount of responsibility for guidance and decision-making, with no financial recompense. At the first Steering Committee meeting, members set priorities that included planting the garden, devising education curricula, developing guides for docents and formulating policy. All of these objectives are being achieved.

Through the creation of a master list of plants for the garden, community members directed the shape of the garden for maximum cultural utility. Local tribal elementary students from the Yocha-De-He school participated in the planting.

Steering Committee members contributed to the design of the outdoor classroom. Concern about chemical contamination of materials to be used for basketry, food, medicine and other purposes guided the decision to practice non-herbicidal weed management, which includes repeated cultivation, burning and hand removal.

Everyone benefits

It is important to understand the motivations and rewards attached to each party's participation in the partnership. This is helpful to research planning, design and evaluation. Of the several participants involved in the Garden, three major parties are markedly different from each other. The gravel industry via the Conservancy, the academic research team and the Native American Steering Committee each has their own driving forces and benefits in this collaborative.

Gravel

For the gravel industry, key reasons for involvement are the obligations within the Cache Creek Resources Management Plan (CCRMP) and with the county that mandate involvement and financial support. In exchange for meeting these requirements, they are permitted to continue to mine along Cache Creek. During a presentation to environmental professionals from around the world, one gravel industry leader told the group that industry did not initially come to the table willingly; but he said they soon learned that it was the right thing to do and these industry leaders are glad they did. Moreover, no one wants to return to the days of the 'Gravel Wars', which damaged public relations for the industry and also hurt personal relationships with friends and neighbors. Local people – homeowners, farmers, county officials, industry representatives and environmentalists – serve on the Conservancy board. Industry officials know that concerns and ideas are coming directly from the community. The industry gets the rare opportunity to be part of the community solution to longstanding problems. As part of a collaborative, industry participants have access to different talents and perspectives that are helping them to plan more carefully for future endeavors. Mining continues with diminished environmental disruption. The industry receives accolades for environmental improvement because of the restoration work. The work of the Conservancy is examined as a potential model for industry–community partnership elsewhere in the state.

Native Americans

The intertribal Committee is concerned that watershed habitat is not available at a level to sustain tribal cultural practices. In the immediate area, basketry has not continued as a strong tradition. Thus, traditional plant management has not persisted as a strong tradition. Once viable gathering areas are now overgrown, subject to chemical exposure, congested with invasive plants and suffer insect predation. The Garden has become a place to teach traditional techniques in a clean environment.

The benefits for the Committee members include the opportunity to learn from each other. Committee members are individually and collectively multi-talented and each person has completely different traditions. The garden offers them the unique opportunity to understand the thinking and learn the languages of industry and the academy – two segments with which traditional native peoples have often been at odds. Committee members can see that their contributions are of benefit to everybody and not just an intellectual exercise. There is access to a venue for balanced public education from a native perspective. There is the considerable satisfaction that comes from creating a project that will endure beyond one's own lifetime.

Researchers

Historically, anthropologists and linguists studied California native tribes intensively. Researchers extracted knowledge, human remains, ceremonial regalia and other cultural material from numerous tribal communities. Universities became repositories and are seen by tribal people as being largely inaccessible. Collectively held, tribal intellectual property was not respected by many researchers. Contemporary critique from various native scholars notes a repeat of this pattern in some research on native management that was purported to be participatory (Simpson, 1999; Simpson, 2000). Community members were not included as colleagues. They did not have a hand in directing the research and did not benefit from it.

In contrast, when time and effort are invested in nurturing relationships of trust, respect and reciprocity, PAR gives academic validity to community-directed research and vice versa. The effort is made to understand what a community holds as valuable and important. This process recognizes the vast body of knowledge owned by Native American communities. The Garden is a manifestation of what can happen when this richness is willingly shared with academic institutions and the broader community.

Two big questions are often asked of Brawley:

1 What is the benefit of this project to the academy?
2 How are your results different than they would have been if you hadn't conducted participatory research?

Brawley was a student at a prestigious public land grant university, whose purpose is to advance knowledge. A quote from the University of California, Davis, website states: 'UC Davis is committed to the tradition of the land-grant university, the basis of its founding. This tradition [is] built on the premise that the broad purpose of a university is service to people and society' (University of California at Davis, 2008). The Garden is an example of research that serves this purpose and therefore produces a positive image of the university to the broader community. In terms of advancing relationships, this project continues to foster a link with one of the university's major benefactors – the Rumsey Band of the Wintun – by providing their children and the Native American community with a traditional educational forum. This project encouraged contributions from a broad section of the philanthropic community that supported on-the-ground research. This created positive publicity for the university. Has this research changed the academy's perceptions and utility of PAR? In a small way, yes. Two professors who advised Brawley on this project now teach PAR in the geography graduate methodology course at UC Davis. As more students, faculty and community members are exposed to this approach, they may chose to incorporate PAR in the research process.

Are the results different than they would have been without PAR? Yes. This research project has fostered relationships between a broad range of community members that might not have come to the table together otherwise. Each stakeholder had a voice in the project. Without this relationship it would just be another area dedicated to a researcher's study of Native American life ways.

Conclusion

When the disenfranchised are invited respectfully to a collaborative research project and treated as colleagues, the results can be amazing. PAR has helped to produce a usable structure for dialogue and the inclusion of traditional indigenous environmental management. The collaborative investigation into the lack of acceptable plant material for native weavers has produced a unique community-designed restoration project. It may also prove to be a social model that will replicate well in other communities.

The Garden is an important and unique contribution that harmonizes beautifully with the vision of the Conservancy. Rather than being absorbed into the overall structure of the Conservancy, the Committee maintains a pronounced identity. The inclusion of the Garden in the CCNP adds the voice of native people to the environmental restoration work in the Cache Creek watershed. Local county agencies have attended restoration workshops held within the Garden. These agencies are working to include within their restoration plans culturally significant plants, traditional management techniques and outreach to the local Native American community who are seeking places in which to gather. The traditional management tools shared by the Committee directly contribute to the Conservancy's mission of wise management, conservation and restoration of habitat. Moreover, the Committee has increased the community capacity for healing after years of discordant histories.

The Garden provides weavers and their students with a starting place to teach and learn traditional management and plant identification, and to gather resources. It has evolved into a natural resource gallery restored with a rich and diverse local riparian plant palette. Moreover, the project provides information for Native American practitioners and others who want to implement similar projects since both the Committee and the Conservancy will have a comprehensive final report on the project. The Garden is still, after several years, one of the CCNP's high-profile projects and a tangible result of participatory research. It provides a venue for sharing cultural knowledge while respecting tribal ownership and has grown into a truly cross-cultural collaborative effort. PAR has created common ground for community members to contribute their expertise where they were once excluded from the process.

REFERENCES

Anderson, K. M. (2005) *Tending the Wild*, University of California, Berkeley, CA

Barbour, M., Pavlik, B., Drysdale, F. and Lindstrom, S. (1993) *California's Changing Landscapes: Diversity and Conservation of California Vegetation*, California Native Plant Society, Sacramento, CA

Blackburn, T. C. and Anderson, K. (1993) *Before the Wilderness: Environmental Management by Native Californians*, Ballena, Menlo Park, CA

Cache Creek Conservancy Board of Directors (1999) *Cache Creek Conservancy Bylaws*, County of Yolo, Woodland, CA

Dianda, M. and Wilson-Wallis, E. (1981) 'Indians' return stirs Capay Valley protest', *Daily Democrat*, 13 February

Greenwood, D. J. and Levin, M. (1998) *Introduction to Action Research: Social Research for Social Change*, Sage, Thousand Oaks, CA

Grenier, L. (1998) *Working With Indigenous Knowledge: A Guide for Researchers*, International Development Research Centre, Ottawa, Canada

Kindon, S., Pain, R. and Kesby, M. (2007) 'Introduction: Connecting people, participation and place', in S. Kindon, R. Pain and M. Kesby (eds) *Participation Action Research Methods and Approaches: Connecting People, Participation and Place*, Routledge, London

Nabhan, G. P. (1997) *Cultures of Habitat*, Counterpoint, Washington, DC

Ortiz, B. (1993) 'Contemporary California Indian basket weavers and the environment', in T. Blackburn and K. Anderson (eds) *Before the Wilderness: Environmental Management by Native Californians*, Ballena, Menlo Park, CA

Reason, P. and Bradbury, H. (eds) (2006) *Handbook of Action Research*, Sage, London

Simpson, L. (1999) *The Construction of Traditional Ecological Knowledge: Issues, Implications and Insights*, PhD thesis, University of Manitoba, Manitoba, Canada

Simpson, L. (2000) 'Aboriginal peoples and knowledge: Decolonizing our process', *The Canadian Journal of Native Studies*, vol XXI, pp137–148

Turner, N. J. (2005) *The Earth's Blanket: Traditional Teachings for Sustainable Living*, Douglas and McIntyre, Vancouver, British Columbia, Canada

University of California at Davis (2007) *Mission Statement: A Philosophy of Purpose*, UC Davis website, http://chancellor.ucdavis.edu/resource/commun/2000/philosophy ofpurpose.cfm, accessed 4 January 2008

Wilmsen, C. (forthcoming) 'Extraction, empowerment and relationships in the practice of participatory research', in B. Boog, J. Preece, M. Slagter and J. Zeelen (eds) *Towards Quality Improvement in Action Research: Developing Ethics and Standards*, Sense Publishers, Rotterdam

6

Opportunities and Challenges in Community Capacity-building: Lessons from Participatory Research in Macon County, North Carolina[1]

Gabriel Cumming, Stacy J. Guffey and Carla Norwood

INTRODUCTION

In Macon County, North Carolina, informal conversations at the bank, church or coffee shop often turn to the rapid growth that the area is experiencing. This topic comes readily to mind because the signs of population influx and new development are everywhere: more homes, more traffic and more outdoor lights creeping up the hillsides. Like many other communities throughout the Southern Appalachian region, Macon County is experiencing unprecedented cultural and ecological transformation. Farms and forests are subdivided for low-density housing even as amenity migrants, such as retirees and second homebuyers, are attracted to the beautiful, rural mountain landscape (McGranahan, 1999). These changes are easy for individuals to remark upon casually; but responding to them as a community has proven much more difficult. Formal discussions of land-use planning options that could shape this rapid growth have been quite contentious, and the minimal land-use regulations that have been enacted give the community little control over the pace and pattern of development.

Little Tennessee Perspectives (LTP), the community-driven initiative described in this chapter emerged in response to, first, the failure of non-participatory planning processes to foster a productive civic dialogue around land-use issues; and, second, the failure of conventional academic research to inform communities' land-use decision-making effectively. LTP represented an attempt to remedy these failures using a self-evaluative, participatory approach.

We begin the chapter by introducing the site and circumstances in which LTP was conceived. Then we describe the methodology and immediate results of the project during 2004 to 2005. The next section focuses on the ongoing project evaluation process. Finally, we reflect on what the research/evaluation experience has taught us about the pitfalls and potentialities of participatory research (PR) as a tool to engage residents in discussing and shaping the future of their local landscapes.

PROJECT SITE: MACON COUNTY, NORTH CAROLINA

Located in the southwest corner of North Carolina, Macon County is nestled within the Southern Appalachians, some of the oldest mountains in the world. Like the arms of a mother, the Cowee and Nantahala ranges surround the Little Tennessee River, which drains most of the county's land area and provides fertile bottomland (LTWA, 2003; see Figure 6.1). This 48km (30 mile) long stretch of valley, where the county seat, Franklin, is located, is flanked on both sides by high plateaus. To the southeast rises the Highlands Plateau, home to the town of Highlands, an exclusive resort community. To the west is Nantahala, an isolated mountain community that seems a world away from the affluent Highlands. Many of the high ridges in the western half of the county are protected National Forest Service lands. In fact, 46 per cent of Macon County is owned by the federal government (Macon County Tax Department, 2005).

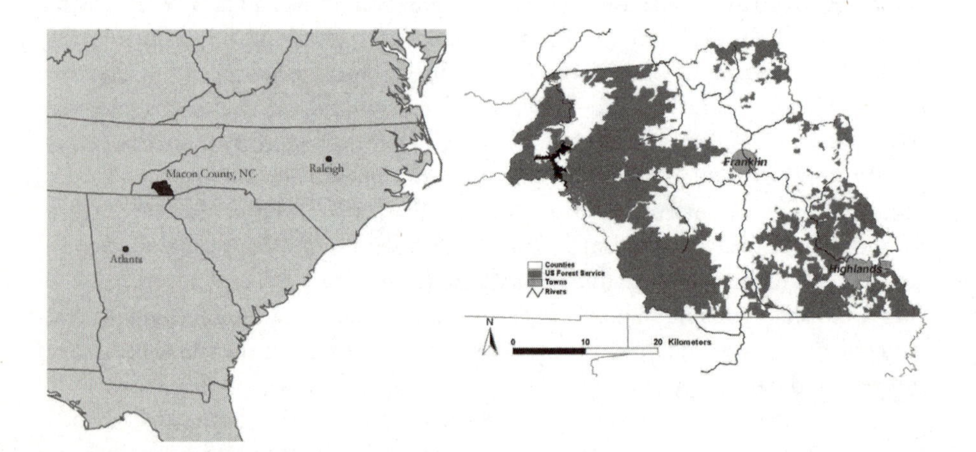

Source: Carla Norwood (2006)

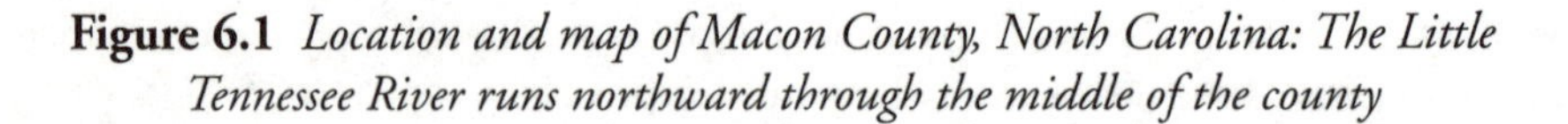

Figure 6.1 *Location and map of Macon County, North Carolina: The Little Tennessee River runs northward through the middle of the county*

A natural corridor for wildlife and human migration, the north–south oriented valley bears evidence of human activity dating back some 10,000 years, when nomadic hunters set up temporary camps in their pursuit of game. With the acquisition of agricultural skills, tribes began constructing permanent settlements. Around 1000 AD, the Mississippian people began construction of the mounds that can still be found in the valley today. Around 1550, the Cherokee became the dominant presence in the area and remained so until white settlers arrived during the late 1700s. With the cession of Cherokee lands in an 1819 treaty, the white settlement of the area began in earnest (Frizzell, 1987).

As with most of the Southern Appalachians, Macon County was settled largely by Scotch–Irish immigrants: some fleeing their homeland because of famine, others being forced to emigrate as part of population-reduction programs instituted by the British government. Being in such an isolated area and seeing themselves as disenfranchised, the Scotch–Irish were quick to mix with the Cherokee. The result was a culture unique to the region (Blethen and Wood, 1998).

Until the mid 1960s, Macon County remained fairly isolated; the people in this area were known for their independence and self-sufficiency. At the same time, because of the necessity of mutual aid, communities and families were extremely tight knit. Their lives and livelihoods depended directly upon the land, so they had a close relationship with it and a deep respect for it. During this period, development generally occurred at the foot of the forested hillsides. The floodplain, being fertile and relatively flat, was used for farmland and pasture. The hillsides and mountains were reserved for timbering and hunting. Homes were built close to water sources, on easily accessible land located near pastures, fields and roads. This pattern reflected the necessities of the times.

During this period, Macon County grew slowly and at times even lost population as a steady outflow of young men and women left the area to seek jobs at plants in Atlanta and Charlotte, or in the booming Detroit auto industry. During the 1960s, as access to the area improved and Americans generated more disposable income, the area was 'discovered'. This process of 'discovery' continued unabated, and today Macon County is one of the fastest growing counties in the region. From 1960 to 1990, the population grew from 14,935 to 23,499, and from 1990 to 2005, the population increased by another 8649 (US Census Bureau, 2005).

The typical newcomers were from Florida – either native Floridians or in-migrants to Florida from the urban areas of the northeastern United States. Having originally fled from the densely populated northern urban areas, this latter group now sought refuge from the rapid growth and sometimes oppressive heat of south Florida. That refuge was a brief vacation to the cool mountains of western North Carolina. Once there, they discovered that local land prices were a fraction of those from where they came, and many of the visitors built second homes that they used in the summer or for vacations. For these new homebuyers, an amenity such as a mountain view or river frontage was more valuable than being close to an existing road or to farm land (Gragson and Bolstad, 2006). With money made

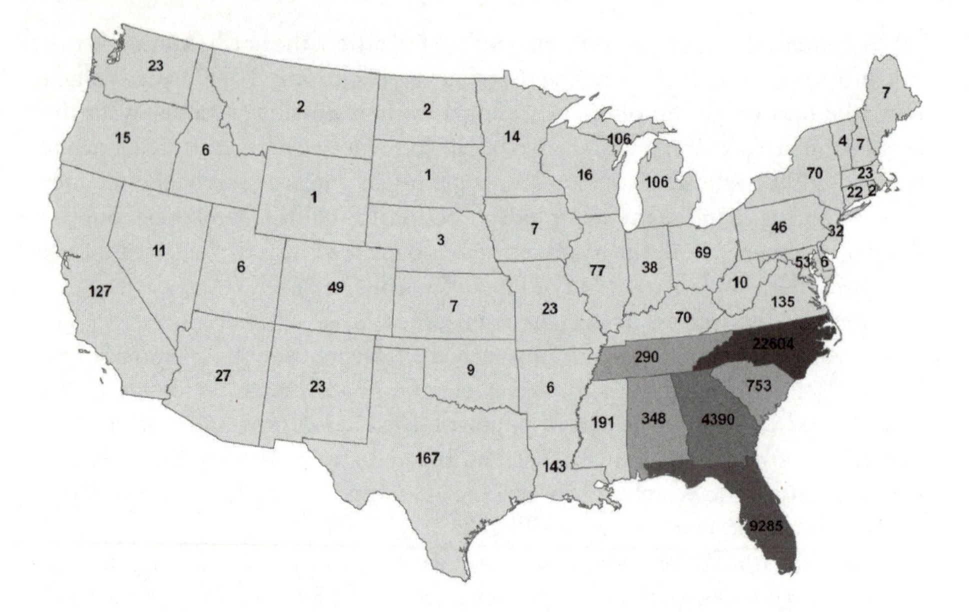

Note: The darker states are home to more Macon County property owners. Forty-three per cent of the county's parcels are owned by out-of-state residents. Twenty-four per cent of the parcels are owned by Floridians, while 10 per cent are owned by people from Georgia. These percentages do not reflect the number of out-of-state in-migrants to Macon County who now list their primary address as North Carolina.

Source: Macon County Tax Department (2005)

Figure 6.2 *Number of property parcels in Macon County owned by people whose primary residence is in each state*

in the booming south Florida real estate market, they were able to build extensive roads to reach the woodlands and high mountains. As longtime landowners realized how desirable their mountainside property had become, many could not resist the lucrative opportunity to subdivide and sell old family land. Local entrepreneurs began real estate and development enterprises.

Today's Macon County is no longer isolated. A newly completed four-lane highway has brought the Atlanta metro area within two hours' driving time. The northern fringes of the Atlanta suburbs creep further north each year. A new wave of in-migrants from the Atlanta area are now purchasing land in Macon County in anticipation of retirement, when many plan to relocate to the area permanently.

The privately owned ridge tops in Macon County, which just 20 to 30 years ago had little monetary value, are now being sold for prices as high as US$0.5 million dollars per acre (US$1.2 million per hectare; Macon County Tax Department, 2005). Each year, more mountainsides are crisscrossed with subdivision roads and

Source: Gabriel Cumming (2005)

Figure 6.3 *The landscape of the Brendletown area in Macon County illustrates the contrast between old and new development patterns: Older houses are located in the valley (**foreground**), while newer houses have been built on the mountainside above (**background**)*

dotted with new homes. The former farms in the fertile floodplains have become prime real estate and are rapidly filling with suburban-style homes. These new development patterns have created many new economic opportunities; but they have also created problems never before faced by Macon County. Rising land values have fostered a booming housing market that is inaccessible to many local working families. Development on the wooded mountainsides has led to runoff, erosion and habitat fragmentation. Steep slope development has raised public safety concerns over slope failures and landslides, while floodplain development has increased the likelihood of property loss and casualties during a flood. These issues have led many community members to join the push for improved land-use planning.

Land-use planning was not on the minds of those who lived in the sparsely populated 'pre-discovery' Macon County; a deep-seated mistrust and resentment of government intervention made the topic taboo. Strong community bonds reduced the need for formal regulation because most disputes could be negotiated informally among neighbors (Rudel, 1989). During the early 1970s, as the real estate boom began and outsiders started settling in the county, planning began to enter into public discussions of the county's future. Draft land-use plans were floated as early as 1974; but serious formal efforts did not begin until the 1990s, when highway corridor protection ordinances, a land-use guidance ordinance and a land-use plan were introduced. In 2002, a comprehensive zoning ordinance was introduced.

However, none of those proposed ordinances, including the zoning ordinance, ever received the political or popular backing necessary to become realities. Each attempt to enact planning measures was met by an organized resistance from a group of property rights advocates who, despite their small numbers, managed to exert significant influence on the public process, often causing public meetings to be quite hostile and rancorous. The methods employed by this group included misinformation campaigns and intimidation. These actions – coupled with a lack of leadership by others – polarized the citizenry. As a result, at most public meetings about land-use issues in Macon County, a few confident speakers have tended to dominate public comment periods, while most people who attend do not participate at all. This antagonistic civic dialogue regarding planning and the failure to reach any agreement that would protect the area's environmental and cultural heritage created a frustrating and negative atmosphere where many residents have felt uncomfortable sharing their perspectives. Except for a sign ordinance, only state or federally mandated ordinances were adopted.

Genesis of Little Tennessee Perspectives (LTP)

In all of the aforementioned Macon County planning efforts, there was a discrepancy between the values most frequently expressed by citizens and the final policy outcome. Conversations held by the authors with a range of residents – from realtors to environmentalists, and from Floridian retirees to long-time natives of the area – suggested that undeveloped vistas, floodplains and ridge tops held value for nearly everyone in Macon County. It seemed possible that a majority of the county's citizens actually supported efforts to protect the landscape from unplanned development, despite the outcomes of the divisive public meetings. A group of local citizens began seeking a fresh approach to talking about these critically important issues before the landscape attributes valued by both current residents and amenity migrants were lost.

The LTP project was conceived through conversations among members of this citizen group and two graduate students at the University of North Carolina (UNC) in Chapel Hill (Cumming and Norwood). The community partners included leaders from the Macon County Planning Board, Macon County government, the Land Trust for the Little Tennessee, a planning advocacy group and a grassroots environmental organization. The project also received the endorsement of the

Macon County Planning Board. One of the students (Norwood) was formerly director of another Franklin-based non-profit organization – the Little Tennessee Watershed Association – and had remained in close contact with colleagues in Macon County after returning to graduate school at UNC. Because of this background, she was able to bridge the roles of 'community member' and 'researcher', thus facilitating the development of a research agenda that was driven by community concerns. Drawing upon the personal experiences of the community members and the student researchers, as well as lessons and methodologies from participatory research, the 'project team' (community partners and student researchers) designed an innovative 14-month project that they hoped would overcome the divisiveness that characterized previous planning debates.

A participatory research approach was seen as offering several advantages. By working collaboratively, community members and student researchers would potentially be able to identify locally appropriate strategies for empowering the community to manage the future of its own land resources. Participatory research may illuminate latent resources that can support empowerment, as well as obstacles that must be overcome. Furthermore, given the widespread distrust of governmental authority in the region, the LTP team believed that top-down approaches to land-use planning were unlikely to succeed; methods that relied fundamentally on the perspectives of residents would perhaps enjoy more popular support (Sanoff, 2000). By providing a model of meaningful citizen participation, the project might be able to challenge the power dynamics that had been inhibiting such participation and lead to more satisfactory land-use outcomes (Greenwood and Levin, 1998; see also Chapter 1 in this volume).

The stated goal of LTP was to foster an inclusive, informed and ongoing conversation about the changing landscape in Macon County – a conversation rooted in community members' shared values. The three components of that goal are worth examining separately. An *inclusive* conversation was considered important because previous planning debates had been so contentious that many residents did not feel comfortable, or even safe, expressing their opinions. The project team wanted to create a forum where everyone felt that their perspectives were valued. Success in advancing this goal would determine whether the perspectives captured in the research were representative of those in the community at large – an issue that has been emphasized in recent critiques of PR (Cooke and Kothari, 2001; Hayward et al, 2004). An *informed* conversation was deemed important, too, because in the past, the public conversation about planning had been hampered by misinformation about policies (sometimes due to deliberate obfuscation) and a general lack of accurate, relevant and accessible data about the rates and long-term consequences of population growth and development. Finally, the team wanted to foster an *ongoing* conversation that would encourage the community to consider issues related to development before crises arose, rather than simply reacting to plans foisted upon them. The project team hoped LTP would serve as a catalyst for a sustained civic dialogue and active citizen engagement around land-use issues. If successful, LTP could thus, ultimately, help build community members' capacity to take a more active role in shaping the future of their shared landscapes (Walker, 2007). Sustainability is one of the principles of community-based participatory research, as defined by Israel et al (2005a); but as Sanoff (2000) points out, follow-up on participatory processes and implementation of ideas generated collaboratively are often overlooked, though critically important, steps.

To achieve these goals, LTP would draw upon local *ecological discourses* – ways in which community members communicate about the environment and their relationship with it (Cantrill and Oravec, 1996; Harré et al, 1999). Success would depend upon the degree to which the project achieved local discursive *relevance*, becoming a part of the community's continuing conversations about landscape change. Argyris and Schön (1991) have argued that 'action research' necessitates a trade-off between rigor and relevance. As Wilmsen (see Chapter 1 in this volume) has pointed out, however, these two goals are not necessarily inconsistent with one another. Indeed, we propose that they may be seen as one and the same: PR should be evaluated on the *rigor* with which it pursues *relevance*. If PR is not relevant to the problems and aspirations of the communities in which it is carried out, then it is not fulfilling its purpose. LTP, then, provided an opportunity to test participatory research's relevance in helping a community determine the future of its own landscape.

PROJECT METHODS

LTP pioneered an iterative, interdisciplinary, participatory methodology for facilitating public dialogue regarding changing communities and landscapes. The three major components of the project – documentary ethnography/video production, analysis and presentation of locally relevant landscape change information, and structured public meetings – were each designed to build on information learned in earlier research stages and to address the root causes of the contentious debates about planning (this methodology is discussed in greater detail in Cumming and Norwood, in preparation).

LTP's research design relied fundamentally on guidance by the community. Over the course of the project, the team sought input on data and methods from increasing numbers of Macon residents, enabling progressive refinement of analyses. Repeated community feedback served as an internal evaluation mechanism for the project: by presenting the research to previously uninvolved local audiences, the project team was able to ascertain whether the research endeavor was staying true to community concerns and discourses. This self-correcting approach is similar to adaptive management frameworks that have been adopted in other natural resource management contexts (Lee, 1999; Berkes, 2004).

LTP's commitment to community needs and perspectives began with the project team's original conversations on problem definition and overall project design. Together, team members discussed previous planning efforts in the area, a variety of potential interventions, the current political landscape, issues of temporal and spatial scale, and how LTP might complement simultaneous or ongoing efforts by other community organizations. This allowed project goals, timelines and research questions to respond uniquely to the local context. Following this important phase, the collection of both ethnographic data and spatial data, which would eventually be shared with the public at a series of meetings, began.

Initial ethnographic data collection consisted of 50 semi-structured interviews with full-time residents of Macon County, who were identified through a combination of snowball and purposive sampling (Bernard, 2002). Interviewees were selected because they were recommended by their peers as having a strong sense of place or an important perspective on the changing landscape. They represented a wide range of perspectives on land-use planning issues, and they closely mirrored the composition of the community in terms of age, gender, race and local/newcomer status. Interviews were conducted both by the student researchers and by six trained volunteers from Macon County; by involving community members in the data-gathering process, the LTP team aimed to increase local ownership of the project and benefit from a diversity of interviewer perspectives (McQuistan et al, 2005). The interviews were audio-recorded, and interviewees were photographed in places that held special significance for them.

Through inductive coding (Patton, 2002) of the transcribed interviews, the student researchers identified emergent *narratives* – shared stories through which the interviewees were positioning themselves in local discourses about landscape and community (Hajer, 1995; Rappaport, 2000). These narratives formed the basis of a 30-minute audio-visual documentary. The documentary, *Macon County Voices*, represents local perspectives on the changing landscape entirely through the words of Macon residents, combining audio excerpts from the interviews with photographs of the interviewees or the local landscape. As will be discussed below, the documentary was reviewed both by community partners and by focus groups to ensure that it accurately and coherently reconstructed the perspectives of Macon residents.

The second major component of the research was to create a clear and accessible presentation that summarized data on the changing landscape. This was intended to provide a more quantitative view of the changes addressed in the documentary and to help construct a foundation of accurate information which the community could reference. Population, land cover and geospatial data were collected from local, state and federal sources. Several advanced spatial analyses responded specifically to the issues most cited by interviewees. For example, many interviewees expressed serious concern about the aesthetic effects of increased homebuilding on mountainsides and ridge tops. Based on this input, the student researchers conducted a community 'viewshed' analysis, highlighting areas of the county that were most visible from the roads and then studying development patterns in the most visible areas (Dean and Lizarraga-Blackard, 2007).

After initial drafts of the documentary and the land-use change presentation were completed, revisions to optimize clarity and accessibility were completed based on comments from community partners and participants in a series of focus groups. During the same period, the project team was also working to plan the public meetings: the third major component of LTP. The goal was to create a more inclusive, reflective and respectful environment in which community members could discuss land-use issues. To accomplish these goals, publicity, as well as the timing, location and internal structure of the meetings were carefully considered. The meetings, which took place on weekday evenings during August 2005, were held in four different parts of the county – Franklin, Cowee, Nantahala and Highlands – in order to maximize the number of residents who could join the conversation.

The meetings, which were conducted by professionally trained facilitators, proceeded according to the following format. Cumming and Norwood presented the land-use change data and screened the documentary *Macon County Voices*. Then small-group discussions started immediately at each table, which prevented anyone from standing up and expressing sentiments that would color all the conversations. Small group facilitators (trained community members stationed at each table) encouraged everyone at their tables to respond to a series of questions, culminating with 'What would you like to see happen to Macon County in the future?' Answers to this last question were brought back to the full group, and all visions for the future were recorded and summarized. At this point, discussion was opened up to the full group. Finally, participants received blank cards on which they could write comments directly to the county commissioners, as well as written evaluation forms seeking feedback on the presentations and meeting structure. A written report, which summarized the project methodology, visions generated and evaluation results, was mailed out to all project participants following the completion of the public meetings.

Initial results of the participatory research

Across the 4 meetings, 127 visions for Macon County's future emerged from the small group discussions. Working with a community partner, the student researchers grouped the 127 visions by topic. Ninety-five per cent of the visions favored collective action to address changes in the county; only 5 per cent advocated doing nothing or upholding the paramount importance of property rights. The most common topics included increased and improved planning, protecting water quality and protecting ridge tops from development.

Written evaluation results suggested that the meeting format was effective – it encouraged participation, introduced relevant information and engendered meaningful dialogue about land-use issues. Notably, 98 per cent of respondents indicated that simply participating in LTP's meeting process increased their support for an ongoing community dialogue around land-use issues. This degree of enthusiasm for a public process, along with the strongly pro-planning sentiments expressed both in the small group discussions and evaluation forms, are unprecedented in the history of Macon County public meetings. The project team was surprised at this deviation from the norm and regarded it as a testament to the significance of the design of a public process. A carefully moderated, participatory meeting and a combative hearing can produce deeply divergent results, even in the same community.

EVALUATING LTP

Despite the unmistakable enthusiasm of LTP participants, the mandate for land-use planning that emerged from the project did not readily translate into increased community capacity to guide landscape change. In the year following the August 2005 public meetings, progress towards a more inclusive, informed and ongoing civic dialogue on land use was mixed at best. In response to these challenges, the project team decided during the summer of 2006 to begin a process of project evaluation. As Israel et al (2005b) observe, evaluation should be an integral element of PR and should be participatory itself. Indeed, the iterative methodology employed in LTP had incorporated evaluation throughout. Nonetheless, a post-project evaluation was warranted in order to reflect on the work thus far and to reposition ongoing efforts. Towards these dual ends, the team determined to pursue two successive lines of enquiry:

1. To what degree has LTP demonstrated its local discursive relevance by fostering increased natural resource management capacity in the community, showing signs of doing so, or failing to do so? Why?
2. To what degree does the project have broader potential relevance to the community that has not yet been realized?

Examining first *demonstrated relevance* and then *potential relevance* reflected recognition that the evaluation process is both backward looking and forward looking: the LTP team wanted to reflect on what had already happened in order to learn from it, and then to consider whether the lessons of the past could be parlayed into greater future effectiveness. The methods used and findings yielded in the course of each enquiry are addressed below.

Looking backward: Evaluating demonstrated relevance

To assess LTP's success in helping build community resource management capacity, the team documented the discernible outcomes of the project since August 2005. This involved tracking public actions that followed directly from the project, actions by project participants and evidence of the project's influence on public discourse. These data were gathered from local media coverage, participant observation and the testimony of community members. The student researchers also conducted follow-up interviews with community partners to elicit their feedback on the project's impact and prospects for next steps.

This investigation revealed positive results as well as missed opportunities. Some positive short-term effects of the LTP process on the public discourse were evident after the meetings concluded. For example, a public hearing on a proposed ordinance to control particularly noxious land uses was dominated by people who had recently attended the LTP meetings. In a marked departure from previous planning ordinance hearings in Macon County, the vast majority of speakers (77 per cent) advocated for increased community action to protect landscape and cultural assets.

The project also advanced dialogue about planning issues in the community and in some ways elevated standards and expectations regarding participation. LTP's effect on public discourse was reflected in, and enhanced by, a high profile in the local media. The project was the focus of at least 17 newspaper articles and editorials in at least seven newspapers in 2005, as well as at least 2 stories on local radio. The media coverage included previews of the documentary, meeting announcements, reports on the meetings themselves and highlights from the project summary report. Evidence from the media suggests that the effects of LTP's 'discursive intervention' in Macon County extended beyond direct discussion of the project. The information about the changing landscape that was co-generated by student researchers and community members has continued to be referenced, including in three newspaper editorials, reflecting the salience of data tailored to community concerns. Landscape terminology introduced during the project, such as *viewshed*, has entered the community lexicon.

Despite LTP's achievement in bringing community members together to discuss shared values regarding the landscape, efforts to maintain the momentum established by those discussions were less successful. Although issues that the project had highlighted came to the forefront of planning discussions following the public meetings, the project has yet to have a significant effect on policy or land-use outcomes. Project participants' desire to enact their visions was frustrated by the response of local policy-makers – or rather, the lack thereof. Most of the county commissioners simply ignored the project, failing to acknowledge the mandates they were receiving from participants. While several members of the county planning board expressed support for the project, a few others found the research threatening and publicly criticized it.

Policy-makers' abortive response to the concerns raised by LTP is exemplified by the steps taken regarding steep slope protections. Following the LTP public meetings, concerns voiced and data gathered through the project propelled the protection of steep slopes to the forefront of the planning board's agenda. A subcommittee was formed to consider enacting rules for development on steep slopes; but a majority of the appointees were affiliated with the development or real estate industries. At the first meeting of the subcommittee, this majority voted to disband themselves, deciding that steep slope protections did not merit consideration (Lewis, 2006).

Decisions such as these signaled a shift of the discourse back towards polarized debate and away from the sense of possibility and collaboration that had accompanied LTP. Project participants found themselves without another clear avenue for ongoing participation, and project team members were discouraged. Participant energy experienced a short resurgence six months later at a forum that project team members organized in response to showcase examples of how other communities are dealing with growth issues. This one-time follow-up event was unable to sustain participatory momentum, however. Public planning meetings turned vitriolic again, alienating many LTP participants who had been inspired to participate in the planning process.

In retrospect, the LTP team realized that it had not adequately planned for the sustainability of the project, and had neglected to focus adequately on the transition between the end of the PR project and the initiation of subsequent steps. Prior to the August 2005 public meetings, the student researchers and community partners were actively engaged with each other and with the community. Much energy was put forward towards a clear and achievable goal that enjoyed popular and political support. The interviews and the public meetings managed to maintain momentum and keep people active and focused on the issue of land-use planning. Although the community partners on the project team played key roles in shaping and carrying out the project, no single leader or organization was responsible for carrying the project forward after the public meetings. There was no obvious point of contact for those interested in staying involved. Perhaps more time clarifying the role of community members and student researchers, particularly for the period following the public meetings, would have helped to avoid this problem.

Looking forward: Evaluating potential relevance

Retrospective study of the year following the LTP meetings revealed that the project's contributions to community capacity-building had been limited both by a failure to thoroughly plan for project sustainability and by a lack of perceived relevance to the ongoing planning dialogue. Community leaders who were inclined to dismiss the project's findings had been able to frame the participants and their views as unrepresentative of the overall community. In evaluating LTP's *potential relevance* to future capacity-building efforts, it would therefore be necessary to assess the *representativeness* of the perspectives in the project.

Evaluating potential relevance provided LTP student researchers with an opportunity to test how effective the project's iterative discursive methodology was at achieving local relevance in the community at large. For the community partners,

Box 6.1 The 'ideal scenario': Community partner Stacy Guffey envisions an alternate outcome

If we were starting the LTP project all over again, what would we do differently? Based on our retrospective evaluation of the project, this is how I think we should have followed up on the public meeting process.

LTP participants, led by the community partners, form a new organization dedicated solely to a campaign for comprehensive long-term land-use planning in Macon County. The organization has a full-time staff member who works to ensure that energy is maintained and focused on the goal. The organization holds regular meetings where community members familiarize themselves with planning and development strategies so that those community members eventually become 'citizen planners'. The organizer ensures turnout at public meetings and other official events to constantly remind decision-makers that long-term planning is the most important issue.

The organization is not adversarial like the anti-planners; rather, it is committed to building working relationships with local leaders. It offers a charismatic, positive vision of Macon County's future. As the organization moves forward in Macon County, staff branch out to help other Southern Appalachian communities replicate its success through projects similar to LTP. These efforts lead to region-wide change. The region becomes a place where citizens take charge of the future of their built environment.

the evaluation became an opportunity to counter the criticisms and dismissals that LTP had encountered in the community. They sought to verify that the narratives, opinions and visions that emerged from the project were resonant not only with participating community members, but also with those who had not participated. If they could secure this affirmation, they hoped to be able to mobilize broader community constituencies around project messages and to command the attention of local governing bodies.

LTP's representativeness was assessed using a sample survey instrument and focus groups. Mail surveys have been rightly criticized as representing only consultative participation (Pretty, 1995; see also Chapter 1 in this volume); but, unlike most social science surveys, this one was embedded within a PR framework: it was developed with community members in response to community demand. The community partners felt that only a scientific sample survey could lend LTP the political credibility that it needed to advance community resource management objectives. Student researchers and community partners designed a survey that would measure respondents' reactions to the statements, information and views that were resonant among LTP participants. The questionnaire also included attitudinal questions about a variety of land-use planning options, including those identified in participants' visions from the LTP public meetings. Focus groups enabled more in-depth exploration of the same topics with small groups of Maconians who had not previously participated in the project.

A thorough discussion of the survey and focus group findings is beyond the scope of this chapter; but results strongly suggest that the attitudes and perspectives that emerged from the participatory research process of LTP accurately captured sentiments that are widely shared in the community. The findings confirm that the majority of residents favor efforts to protect valued landscape attributes such as rural character, clean water and forested hillsides. The project team is currently interpreting the data and will be designing an appropriate strategy during the coming months for sharing the results with the community at large and with decision-makers.

Discussion

Evaluating LTP's pursuit of relevance, as measured through representativeness and sustainability, reveals strengths and weaknesses that are both instructive in seeking to improve the benefits of participatory research to communities. This participatory research effort enjoyed greater success in engaging community members in discussions about land-use planning issues than had any previous, less participatory, projects. By all indications, LTP's participatory research process succeeded in identifying perspectives that were representative of those held in the community at large and therefore enjoyed considerable local relevance. Moreover, the August 2005 meetings demonstrated that invoking those perspectives could foster meaningful participation and effective community dialogue: meeting participants were able to carry on successful conversations around land use in a community where the topic is usually divisive, and they gave the meeting format high marks in written evaluations. The project brought together the competencies of both community members and student researchers to foster an environment of possibility, while also helping to deconstruct stereotypes about divisions between 'locals' and 'outsiders' that have often distracted people from addressing shared concerns. The project garnered support and participation from Maconians across the ideological spectrum. The evaluative survey results attest that such apparent solidarity was no illusion; the views, concerns and aspirations of project participants are broadly shared by county residents at large. These findings suggest that, through sensitivity to local discourse, participatory research can generate trustworthy findings and successfully enter and advance community dialogue.

LTP's iterative process also demonstrated how participatory methods can be 'scaled up' to involve ever-larger populations. A core group of community partners and student researchers engaged most deeply and most consistently in designing and carrying out the project. However, as the research and evaluation progressed, increasing numbers of community members became involved. Over the course of the last 3 years, participation expanded from a core group of 6 to a sample survey that reached 1800. Each broadening of the participant pool enabled the project team to review and revise its understanding of salient community concerns and narratives. This nested methodology encouraged participation from a wide variety of residents, not just those who typically participate in discussions of land-use issues (Cooke and Kothari, 2001).

However, regarding sustainability, LTP fares less well. Participatory research projects can be seen as efforts to involve community members in the discursive construction of alternate social realities: realities in which it is possible to shift power dynamics and solve a stubborn problem or rectify an injustice. From this rhetorical perspective, participatory research projects resemble social movements, which DeLuca (1999, p36) describes as 'changes in the meanings of the world, redefinitions of reality' that challenge prevailing societal conceptions of an issue (see also McGee, 1975). LTP had opened a window in the hegemonic local planning discourse to reveal an alternative articulation of discursive elements (DeLuca, 1999) – a more empowering way of framing the issues. In the wake of the August 2005 public meetings, project participants could not sustain the project's energy and hold this window open long enough for this discursive re-articulation to gain currency, and so the divisive debates resumed.

The project team was unable to expand the participatory process enough to encompass Macon County as a whole; the internal discourse of the project could not readily be integrated within Macon County's ongoing political discourse. LTP had encountered a problem that is likely to confront many participatory research efforts: the alternate reality that had been created within the context of the project was actually more participatory than the local political environment. Undertaking an evaluation that consciously employed less participatory methods was needed to translate the messages from the project into this external – and less participatory – discursive space. Based on our experience, we would advise other PR teams to anticipate the need for such translation and plan for it. Developing a clear strategy (or strategies based on two to three likely scenarios) for following up on the project is essential. From the outset, then, participatory research project teams should consider not only how to establish a locally relevant alternate reality, but how to sustain that reality once constructed. Interrogating assumptions about proper roles and tasks for each member of the project team is also necessary because it is likely that community and academic partners will offer different strengths or be limited by different constraints at this stage as well.

Even if a participatory message initiative loses momentum, however, it does not necessarily mean that no progress has been made: ideally, the sense of possibility engendered in the community is obscured but not forgotten. Although the ongoing discussion about land use and values that team members hoped to foster through LTP encountered setbacks, the alternative reality modeled by the project has retained local discursive relevance. More than two years after the public meetings, citizens from Macon County and many other Southern Appalachian counties continue to demonstrate interest in the project, requesting presentations, sharing copies of the documentary and land-use presentation, or asking for updates from the authors. In fact, dissemination of project materials occurs largely without the knowledge of the authors – which pleases us because it suggests that the community has ownership of at least some aspects of the project. The current chair of the Macon County Planning Board has credited the project with advancing dialogue about planning by at least two years. LTP demonstrated, in Macon County and throughout the region, that a higher standard of participation could be achieved. And, unexpectedly, a seven county project aimed at advancing land-use planning in the region has adopted this participatory methodology to identify shared values and perspectives in advance of a regional planning charrette in 2008. Finally, and significantly, the student researchers and the community partners have maintained close and productive relationships.

CONCLUSION

We believe that participatory research techniques such as those described here, although relatively untested, may provide critical methods and insights for addressing landscape change in the rural US. In essence, the sense of helplessness that many rural communities feel when confronting sprawl development reflects a failure of democracy: community members are not empowered to exercise their rights as citizens (Holland et al, 2007). *Disempowerment* is a term typically associated with systematically disenfranchised populations – it may not seem to apply to white middle-class residents of the rural US, for example, who could be considered one of the most politically powerful constituencies in the world. Our research suggests, however, that rural North Carolinians, regardless of race, gender, income or other demographic characteristics, feel profoundly alienated from the decision-making processes that determine future development in their local landscapes. Many community members feel like their only option is to comment informally on the destruction of their landscape; but, by and large, they do not feel empowered to participate in its protection at the community scale. Whether or not these citizens face systemic institutional barriers to participation, their disempowerment is discursively real and therefore functionally real: it is manifest in the development of the biophysical landscape.

Only participatory initiatives that challenge the discursive basis of this disempowerment stand a chance of overcoming it. Researchers, professionals and community activists in the natural resources field, then, face a moral imperative to help communities build their capacity to manage landscape change. However, simply adopting a participatory approach is insufficient. A participatory project should be designed and evaluated according to rigorous criteria for the pursuit of relevance. In our experience, the initial framing of research goals and methods is a critical point for establishing understanding and trust between community and research collaborators. Furthermore, embedded and ongoing evaluation can help to ensure that community concerns are being addressed and that efforts are representative of the constituencies whom they claim to represent. Sustainability should be considered not only in terms of the immediate project goals, but also in broader terms that can allow aspirations and strategies to evolve as communities change and build capacity. When development pressures jeopardize the natural and cultural heritage of the communities in which they work, PR practitioners should do their part in helpimg to address those pressures.

ACKNOWLEDGMENTS

The authors would like to thank all the Macon County residents – more than 1100 in all – who have contributed their time and insights to this project. In particular, we are very grateful to the other members of the LTP project team – Ben Brown, Dennis Desmond and Susan Ervin – who have been our guides and collaborators throughout. This research was made possible through financial support from the Community Forestry Research Fellowship Program, Community Foundation of Western North Carolina, Highlands Biological Station, Morris K. Udall Scholarship and Excellence in Environmental Policy Foundation, National Science Foundation Graduate Research Fellowship Program, and the Royster and University Fellowship Programs (Graduate School, University of North Carolina at Chapel Hill).

NOTES

1 Portions of this chapter also appear in Cumming (2007).

REFERENCES

Argyris, C. and Schön, D. A. (1991) 'Participatory action research and action science compared: a commentary', in W. F. Whyte (ed) *Participatory Action Research*, Sage Publications, Newbury Park, CA

Berkes, F. (2004) 'Rethinking community-based conservation', *Conservation Biology*, vol 18, no 4, pp621–630

Bernard, H. R. (2002) *Research Methods in Anthropology: Qualitative and Quantitative Approaches,* 3rd edition, AltaMira Press, Walnut Creek, CA

Blethen, H. T. and Wood, Jr, C. W. (1998) *From Ulster to Carolina: The Migration of the Scotch-Irish to Southwestern North Carolina*, North Carolina Department of Cultural Resources Division of Archives and History, Raleigh, NC

Cantrill, J. G. and Oravec, C. L. (1996) 'Introduction' in J. G. Cantrill and C. L. Oravec (eds) *The Symbolic Earth: Discourse and Our Creation of the Environment*, University Press of Kentucky, Lexington, KY

Cooke, B. and Kothari, U. (eds) (2001) *Participation: The New Tyranny*, Zed Books, London

Cumming, G. (2007) *Explorations in Discursive Ecology: Addressing Landscape Change with Rural North Carolinians*, PhD thesis, University of North Carolina at Chapel Hill, NC

Cumming, G. and Norwood, C. (in preparation) 'An iterative, participatory method for involving rural communities in conversations about changing landscapes'

Dean, D. J. and Lizarraga-Blackard, A.C. (2007) 'Modeling the magnitude and spatial distribution of aesthetic impacts', *Environment and Planning B: Planning and Design*, vol 34, no 1, pp121–138

DeLuca, K. M. (1999) *Image Politics: The New Rhetoric of Environmental Activism*, Guilford Press, New York, NY

Frizzell, G. E. (1987) 'The Cherokee Indians of Macon County', in J. Sutton (ed) *The Heritage of Macon County, North Carolina*, Macon County Historical Society, Winston-Salem, NC

Gragson, T. L. and Bolstad, P. V. (2006) 'Land use legacies and the future of Southern Appalachia', *Society and Natural Resources*, vol 19, pp175–190

Greenwood, D. J. and Levin, M. (1998) *Introduction to Action Research: Social Research for Social Change*, Sage Publications, Thousand Oaks, CA

Hajer, M. A. (1995) *The Politics of Environmental Discourse*, Clarendon, Oxford, UK

Harré, R., Brockmeier, J. and Mühlhäusler, P. (1999) *Greenspeak: A Study of Environmental Discourse*, Sage Publications, Thousand Oaks, CA

Hayward, C., Simpson, L. and Wood, L. (2004) 'Still left out in the cold: Problematising participatory research and development', *Sociologia Ruralis*, vol 44, pp95–108

Holland, D., Nonini, D. M., Lutz, C., Bartlett, L., Frederick-McGlathery, M., Guldbrandsen, T. C. and Murillo, Jr, E. G. (2007) *Local Democracy under Siege: Activism, Public Interests, and Private Politics*, New York University Press, New York, NY

Israel, B. A., Eng, E., Schulz, A. J. and Parker, E. A. (2005a) 'Introduction to methods in community-based participatory research for health', in B. A. Israel, E. Eng, A. J. Schulz and E. A. Parker (eds) *Methods in Community-based Participatory Research for Health*, Jossey-Bass, San Francisco, CA

Israel, B. A., Lantz, P. M., McGranaghan, R. J., Kerr, D. L. and Guzman, J. R. (2005b) 'Documentation and evaluation of CBPR partnerships: In-depth interviews and closed-ended questionnaires', in B. A. Israel, E. Eng, A. J. Schulz and E. A. Parker (eds) *Methods in Community-based Participatory Research for Health*, Jossey-Bass, San Francisco, CA

Lee, K. N. (1999) 'Appraising adaptive management', *Conservation Ecology*, vol 3, no 2, www.consecol.org/vol3/iss2/art3, accessed 1 November 2006

Lewis, M. (2006) 'Developers want "steep slope" dead – subcommittee disbands itself', *Franklin Press*, 2 May, www.thefranklinpress.com/articles/2006/05/02/news/01news.txt, accessed 18 March 2008

LTWA (Little Tennessee Watershed Association) (2003) *State of the Streams of the Upper Little Tennessee Watershed*, LTWA, Franklin, NC

Macon County Tax Department (2005) *Macon County Tax and Mapping Site*, Text download, www.maconnc.org, accessed 20 May 2005

McGee, M.C. (1975) 'In search of the "the people": A rhetorical alternative', *Quarterly Journal of Speech*, vol 61, pp235–249

McGranahan, D. A. (1999) *Natural Amenities Drive Rural Population Change*, United States Department of Agriculture Economic Research Service, Washington, DC

McQuiston, C., Parrado, E. A., Olmos-Muñiz, J. C. and Martinez, A. M. B. (2005) 'Community-based participatory research and ethnography: the perfect union', in B. A. Israel, E. Eng, A. J. Schulz and E. A. Parker (eds) *Methods in Community-based Participatory Research for Health*, Jossey-Bass, San Francisco, CA

Patton, M. Q. (2002) *Qualitative Research and Evaluation Methods*, 3rd edition, Sage Publications, Thousand Oaks, CA

Pretty, J. N. (1995) 'Participatory learning for sustainable agriculture', *World Development*, vol 23, no 8, pp1247–1263

Rappaport, J. (2000) 'Community narratives: Tales of terror and joy', *American Journal of Community Psychology*, vol 28 pp1–24

Rudel, T. K. (1989) *Situations and Strategies in American Land-use Planning*, Cambridge University Press, Cambridge, UK

Sanoff, H. (2000) *Community Participation Methods in Design and Planning*, John Wiley and Sons, New York, NY

US Census Bureau (2005) 'Census summary file 3 (SF 3)', Data generated by Carla Norwood, using American FactFinder, http://factfinder.census.gov, accessed 13 May 2005

Walker, G. B. (2007) 'Public participation as participatory community in environmental policy decision-making: From concepts to structured conversations', *Environmental Communication*, vol 1, no 1, pp99–110

7

Calibrating Collaboration: Monitoring and Adaptive Management of the Landscape Working Group Process on the Grand Mesa, Uncompahgre and Gunnison National Forests in Western Colorado

Antony S. Cheng, Kathleen Bond, Carmine Lockwood and Susan Hansen

INTRODUCTION

Collaboration occurs when individuals voluntarily work through their differences and share knowledge and resources to achieve goals that they could not achieve alone (adapted from Gray, 1989, and Daniels and Walker, 2001). Developing long-term management plans for public forests in the US is a context in which collaboration among stakeholders is essential. The issues are numerous, frequently interconnected and defy the capacity of any one entity to address holistically (Daniels and Walker, 2001). Participants come with incomplete knowledge and hold values and expectations that are not always well defined and/or conflict with one another. However, studies have shown that through an adaptive collaborative process, participants can engage in social learning about the issues and one another's perspectives, thereby facilitating the development of shared knowledge of, and goals for, the resources in question (Daniels and Walker,1996; Blatner et al, 2003; Bouwen and Taillieu, 2004; Schusler et al, 2003).

This chapter analyzes the strategies for making a collaborative process adaptive so that it enables participants to develop a national forest plan that is mutually acceptable and feasible to implement. The strategies focus on the development, measurement and evaluation of monitoring criteria and indicators for the Landscape Working Group (LWG) process that occurred during the revision of the Grand Mesa, Uncompahgre and Gunnison National Forests (GMUG) forest plan. The adaptive management of a collaborative process parallels the adaptive management framework grounded in ecosystem management (Holling, 1978; Walters, 1986), in which management actions are treated as experiments from which learning is a critical product. Learning is achieved through the systematic monitoring of key indicators and critical evaluation based on established criteria of performance. In turn, learning leads to a refinement of management actions or even to the termination of action. This learning loop is most effective when multiple perspectives are involved; the integration of multiple perspectives can lead to more effective and sustainable outcomes (Hemmati, 2002).

The lenses through which this analysis occurs are the perspectives of three individuals who served on an *ad hoc* 'collaboration team' that continually monitored, evaluated and modified the LWG process, as well as a community leader who was an active participant in several LWGs. The team includes Carmine Lockwood, forest planning staff officer and team leader for the GMUG forest plan revision process; Kathy Bond, under contract to the US Institute for Environmental Conflict Resolution as the lead facilitator of the LWG process; and Tony Cheng, associate professor at Colorado State University contracted through the United States Forest Service (USFS) Rocky Mountain Research Station to provide assistance in the design, implementation and evaluation of the collaborative public involvement in the GMUG plan revision. Susan Hansen, Delta County administrator and a rancher in Crawford, Colorado, represented the interests and concerns of Delta County in the Uncompahgre and North Fork Valley LWGs, as well as her personal perspectives as a rancher in the North Fork Valley.

This type of participatory adaptive process management mirrors the collaboration itself and presents numerous opportunities for improving meaningful community involvement in public forest planning and management. It is also fraught with challenges because each collaboration team member had specific roles, responsibilities, personal expectations and professional objectives in the LWG process. As a consequence, each team member focused on certain monitoring criteria, indicators and mechanisms more than others.

The chapter is organized into four parts. The first provides the contextual background for the GMUG forest plan revision and LWGs. The second part describes the monitoring criteria, indicators and mechanisms, and the opportunities for adaptive management of the LWG process. Integral to defining monitoring criteria are the expectations, objectives and biases that each collaboration team member brought to the effort, which, in turn, generated a creative tension around the adaptive process management. This is followed by a description of what information the indicators generated, how information and feedback were weighted, what process changes were made as a result of monitoring, and the effectiveness of those process changes. The final section is a discussion about lessons

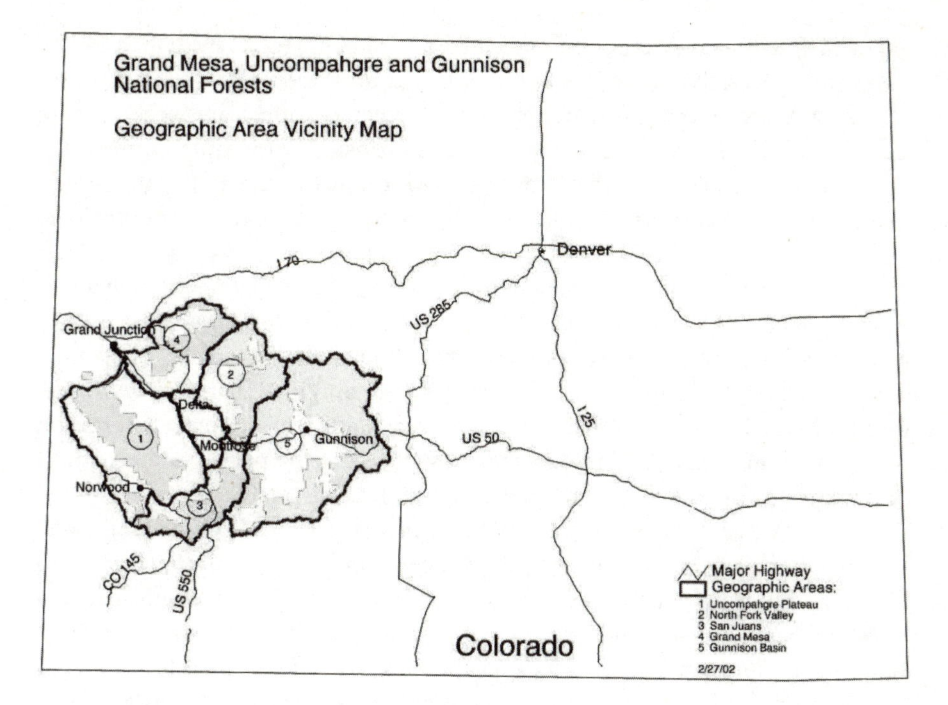

Source: GMUG Planning Team (February, 2002)

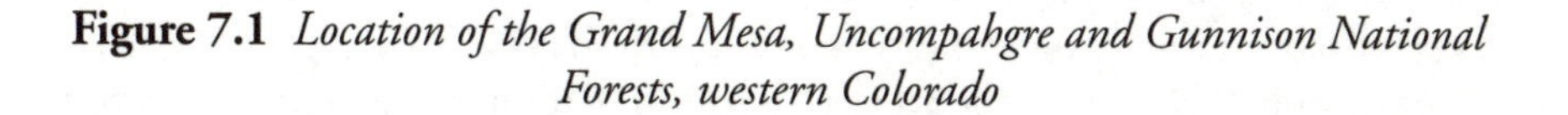

Figure 7.1 *Location of the Grand Mesa, Uncompahgre and Gunnison National Forests, western Colorado*

learned as a team and how these lessons can be applied to other collaborative processes in public forest planning and management.

Contextual background

The GMUG encompasses approximately 1.2 million hectares (3 million acres) in Western Colorado and is managed according to a forest plan, a document required under the 1976 National Forest Management Act that defines forest management goals, priorities and objectives for 10 to 15 years. In 1999, the process to revise and update the GMUG plan was initiated by the forest's leadership (forest supervisor and district rangers). The first phase of the GMUG forest plan revision was to develop assessments for each of the five geographic areas comprising the GMUG: the Uncompahgre Plateau, the North Fork Valley, the Grand Mesa, the San Juans and the Gunnison Basin. In general, an assessment compiles information on key issues and concerns, current conditions, historic and/or reference conditions, trends and risks, and desired future conditions for a geographic area. As such, the GMUG assessments generated the core decision elements for the revised forest plan.

The LWG concept was conceived by the collaboration team working with the GMUG Core Planning Team[1] for the purpose of providing community stakeholders and other interested publics with opportunities to collaboratively define desired conditions and management goals for the five geographic areas comprising the GMUG. The LWGs, based loosely on Daniels and Walker's (2001) collaborative learning approach, sought inclusive, open participation through media advertisements, invitations to individuals on the GMUG's regular mailing list and targeted recruitment by email or phone of representatives of organized stakeholder groups, local elected officials, tribal representatives and unaffiliated citizens. The organized stakeholder groups included motorized and non-motorized recreation users, environmentalists and resource users, such as grazing permit holders, forest products industries and fossil fuel industries (natural gas and coal).

An LWG was not a distinct entity but a mobile event; LWG meetings were held in different communities within each geographic area and, therefore, had different attendees. The purpose of the LWG meetings was for USFS staff and public stakeholders to collaboratively develop desired conditions and management recommendations for the GMUG's geographic areas. Between February 2002 and October 2003, 42 community meetings were conducted during weekday evenings across the 5 areas with 1620 participants according to the meeting sign-in sheets. Fifteen per cent of the participants were formally affiliated with an organized interest group; the remainder did not claim an organizational or interest group affiliation. Due to the continuous monitoring and adaptations conducted by the collaboration team, the LWGs evolved over the 18 months and changed as the process moved – west to east – from the Uncompahgre Plateau to the Gunnison Basin. The following two sections describe the specific monitoring components and explain how monitoring feedback led to specific collaborative process adaptations.

APPROACH TO MONITORING AND ADAPTIVELY MANAGING THE COLLABORATIVE PROCESS

The basic strategy taken by the collaboration team was to constantly refine LWG meetings to find the best fit between the substantive contents of the assessments (i.e. desired conditions and management recommendations), LWG participants' sophistication and expectations for how they wanted to influence the forest planning process and the capacity of, and constraints on, the collaboration and core planning teams to carry out the collaborative learning process. Drawing on the adaptive ecosystem management literature, the LWG process monitoring contained three core components: criteria, indicators and measurement mechanisms (see Margoluis and Salafsky, 1998; Wright and Colby, 2002). Criteria specify what is important and what goals are desired from a given enterprise. Indicators are discrete metrics or data points that provide information about the level of performance. Measurement mechanisms are the various techniques used to acquire indicators.

Criteria

Daniels and Walker (2001) define three criteria for making progress within a collaborative process: substantive, procedural and relationship. The *substantive* criterion was defined as being attained when the assessment documents were completed and when all stakeholders viewed the assessments as accurate, rigorous and fair. The *procedural* criterion was defined as being attained if participating and non-participating stakeholders perceived that the LWG process included broad stakeholder participation; the open exchange of ideas and information; accurate ecological, economic and social information; and equal time and opportunity for dissenting opinions to be voiced. Lastly, the *relationship* criterion was defined as being attained when working stakeholders involved in the LWG process perceived their participation to be characterized by positive interactions, open communication, mutual learning, respectful listening and consideration for other perspectives.

Each collaboration team member had their own assumptions and expectations for which criterion was most important. For Carmine: 'My focus was on making improvements in the relationships dimension, rather than resolving complex and controversial resource issues, which is what I'd characterize more as the substance dimension.' Improving communication and trust between the USFS and community stakeholders throughout the plan revision and into implementation was paramount. As the professional third-party facilitator, Kathy saw herself as 'an advocate of the collaborative process' and had a responsibility to:

> *... interact with interested individuals and interest groups with a stake in the outcomes of the collaborative process. This includes non-participating stakeholders to make sure they perceived that the USFS was incorporating a broad diversity of stakeholder input into the decision-making process and not favoring any one group's viewpoints.*

These sentiments fall squarely in the procedural criterion.

Tony's research interest, which focuses on both the process structure and relationship-building, colored his expectations of the collaborative process: 'The process structure and working relationships have to be established before stakeholders can tackle the substantive issues.' Tony had the expectation that there was somewhat of a sequence to collaboration and intended to measure the relationship between process structure, relationship-building and substantive outcomes. With nearly a decade of working with the Public Lands Partnership, a collaborative group of individuals representing diverse interests in the current and future management of the GMUG, Susan had a pragmatic, integrated vision for the LWGs:

> *As a community, we embraced the concept of a collaborative learning process and the notion of being an active and engaged partner in all aspects of the Forest Plan Revision process as the public lands are very important to the economic, social and environmental health of our rural communities.*

If the collaborative process made a difference in the landscape and in people's lives, it would be a success.

It is important to highlight these different assumptions and expectations because each collaboration team member, as well as active participants such as Susan, emphasized and weighted the information from the monitoring through his or her own particular lens. The collaboration team learned early of the importance of making explicit and clearly articulating these perspectives when it became clear that there were multiple and sometimes competing personal agendas for how the LWG process should proceed.

Indicators

As the collaboration team members came to understand one another's personal and professional expectations and objectives for the collaborative process, they collectively developed indicators for specific types of outcomes. The indicators were both qualitative and quantitative. Qualitative indicators were in the form of oral comments during interviews, meetings, informal conversations or in written comments on post-meeting evaluation questionnaires. Quantitative indicators were derived from pre- and post-working group and post-meeting questionnaires. An attendance record was kept to indicate how many and what types of stakeholders participated in the LWG process.

Substantive indicators included:

- participants' understanding of:
  - the geographic area assessment;
  - the forest plan revision;
  - the range of issues in the geographic area;
  - the potential management options;
  - the information presented at LWG meetings;
- development of a range of desired future conditions;
- identification or focusing of goals, objectives and other substantive elements that the Core Planning Team needs to write the draft assessment;

- participants' (including agency staff) perceptions that substantive objectives are being met.

Procedural indicators included:

- participants' perceptions that:
  - LWG is an open, fair process;
  - the LWG process is achieving substantive goals and objectives;
  - increased learning of issues and options is occurring among participants;
  - participants are working well together;
  - the LWG meeting is a good use of time and energy;
  - the quality of facilitation is high;
- comfort level with activities;
- level and diversity of participation in LWG meetings;
- level of conflict over accuracy of information;
- number of participants attending throughout the LWG process, including returning participants and 'dropouts';
- infrequently participating or non-participating stakeholder perceptions that:
  - the LWG process is non-exclusive and open;
  - the LWG process is not being dominated by one stakeholder interest.

Relationship indicators included:

- participants' perceptions that:
  - participants are working together;
  - there is increased learning about issues and options among participants;
  - there is increased communication among participants;
  - there is increased respectful listening among participants;
  - barriers are being removed between participants;
  - common ground or shared interests exist between diverse participants;
  - new types of relationships are being formed between participants as a result of the LWGs;
- infrequently participating or non-participating stakeholder perceptions that:
  - USFS is making a good faith effort to incorporate diverse issues and input;
  - USFS can be trusted to conduct rigorous analysis of information.

Six types of mechanisms to gather information about indicators were developed: questionnaires; post-LWG meeting debriefings; post-meeting phone interviews with selected LWG participants; feedback from district staff meetings; periodically scheduled Core Planning Team meetings; and *ad hoc* informal conversations with key contacts and LWG participants.

Pre- and post-LWG questionnaires measured changes in stakeholder perceptions, understanding and learning throughout the process. A second questionnaire was distributed to LWG participants after every meeting. The questionnaires generated quantitative measures that tracked participants' feedback from meeting to meeting throughout a working group's entire process, as well as compared feedback across geographic areas. Written comments added depth to participants' ratings of LWG meetings. The questionnaire provided information directly for the substantive and procedural criterion.

At the end of each meeting, Kathy asked participants to identify positive aspects of the meeting and what aspects could be changed. These comments were recorded on a flipchart. Immediately following each meeting, Kathy conducted a debriefing with Core Planning Team members and participating district staff. The debriefings were qualitative and emphasized procedure, but touched on all three criteria. Key questions during the debriefings were: 'What worked well in the meeting for you?' and 'If we were having another meeting in three days, what improvement(s) would you recommend?'

Within three days after an LWG meeting, Kathy conducted phone interviews with five LWG meeting participants or interested stakeholders. The interview was semi-structured and focused on three questions:

1 Is the LWG group moving forward to identify goals/objectives that the Core Planning Team needs in order to write the assessment documents?
2 How is the group working together? What signs are there that barriers are being removed?
3 How is the process design working to provide information as well as to facilitate group interaction and mutual learning?

Kathy recorded the interviews verbatim and distributed summaries to the Core Planning Team within one week of the LWG meeting, kept in an LWG record book. Once every three weeks, on average, the Core Planning Team met to discuss the progress of the geographic area assessments and other forest plan revision work tasks. This was a key mechanism by which the substantive dimensions were addressed in the LWG process. Feedback from the team on the LWG process was qualitative and primarily focused on substance and procedural criteria.

The last feedback mechanism included informal conversations between Kathy, Carmine, members of the Core Planning Team and stakeholders. These conversations were scheduled phone calls or meetings, as well as spontaneous encounters at the post office, grocery store, movie theater, or high school baseball game. Extra effort was made to converse with community leaders, such as elected county and city officials, interest group representatives and other prominent opinion leaders within the community.

INTERPRETING MONITORING FEEDBACK: LESSONS LEARNED AND ADAPTATIONS

Feedback and adaptations relating to the substance criterion

One of the first lessons was that participants needed a clear idea of what plan revision decisions they could actually influence. Participants were also eager for information about forest conditions and issues from the USFS. Stakeholders preferred to respond to USFS information rather than start from scratch and contribute their own information. This came as somewhat of a surprise since the LWG process was initially designed to be more stakeholder driven than agency driven.

A bigger surprise was that stakeholders were much more sophisticated than the core planning and collaboration teams had expected. There was variation in the feedback; but Carmine noticed that: 'They can handle more complexity than we originally thought appropriate for the target audience. They multi-task and multi-concept quite well. They grasped landscape management theme activities well; these are complex exercises dealing with multiple decision elements and assessment components.' The landscape management theme activity was a participatory mapping exercise in which LWG participants deliberated current and desired future conditions using a variety of data and information, including participants' own experiences (see Cheng and Fiero, 2005).

The Core Planning Team developed the landscape management theme activity to facilitate discussion about current and potential future conditions for the landscape unit. Instead of addressing management options for an entire geographic area (approximately 250,000ha to 300,000ha, or 617,500 acres to 740,000 acres), the LWG process focused on smaller landscape units (approximately 10,000ha to 17,000ha, or 24,700 acres to 42,000 acres). These smaller units were at a more 'human scale' that corresponded to participants' experiences and personal knowledge of the land and resources.

The landscape units were delineated by GMUG district rangers and staff and assigned a name commonly recognized by the community, such as Taylor Park or Mechanic Mountain. Each unit was assigned a current condition according to a thematic classification system. Theme 1 are areas dominated by natural processes and are conditions that would make them suitable for wilderness designation, while theme 8 signifies areas permanently altered by human activities, such as ski areas. Theme 5 represents areas with substantial human activities, such as livestock grazing and motorized recreation, but where natural processes still exist (i.e. multiple-use lands). Maps for each of the GMUG's five geographic areas were generated and displayed each landscape unit's name. Each unit was color coded by theme.

For the activity, participants were randomly assigned to teams of six individuals. Each group was equipped with a designated facilitator (usually a district staff member); maps of the geographic area divided into the color-coded landscape units; a landscape theme reference guide; summary information tables for each landscape unit; and a desired condition worksheet. Participants confirmed or amended current conditions and defined desired conditions for each landscape unit per thematic classification system. Participants compared and contrasted their own knowledge and experience on the specific landscape unit and often challenged one another about whether conditions have remained the same, improved or worsened. Comments and changes were recorded directly on the map or on the data summary sheet, including dissenting viewpoints. Each geographic area had between six and nine LWG meetings, depending upon the number of communities in the area, so that participants could rotate through all of the landscape units in the area.

Feedback and adaptations relating to the procedural criterion

In general, the feedback suggested strong support for the collaborative learning approach embodied in the LWG process. Regular participants were invested in the LWG process and wanted to maximize the opportunities for shaping the revised forest plan. Participants also recognized the need to adaptively manage the LWG process and appreciated opportunities to give suggestions for improving the process. On a meeting-to-meeting basis, participants highly valued the opportunity to discuss GMUG forest issues with other participants, even those who hold different views and values. However, there were indications that specific LWG exercises and activities needed to be more clearly explained, especially how each activity related to the overall assessment and plan revision process. They also wanted to see their input reflected in products that could be tracked throughout the plan revision process. As a result, district staff and Core Planning Team members spent time during each LWG meeting on a sample landscape management theme activity before turning participants loose to work on their own. LWG meeting outcomes and products were posted on the GMUG website and were updated as the assessments were completed. LWG products are clearly seen in the draft plan released in March 2007.[2]

There was a broader concern about dwindling communication between the USFS and participants as time went on. Susan observed this decline and its impacts:

> *The lack of communication was not keeping the Forest Plan Revision process in front of the community. The announced time schedules seemed to slip in terms of getting back to the community with 'deliverables' or products. Community members move on to other things and lose interest and/or commitment to the effort if they are not kept in the loop.*

The communication delays were due, in large part, to the Core Planning Team's assumption that the assessments would be developed over the course of the LWGs. When this did not occur, the Core Planning Team had to devote extra time to work on the assessments – time they did not expect to spend. As a result, there were time gaps between LWG meetings as the Core Planning Team worked on the assessments.

One unresolved issue was that the LWG process would produce mutually supported outcomes and working relationships at the local level, but would be ignored by USFS superiors or politicians at the regional and national levels. This concern was confirmed in July 2006 when, just days before the release of the draft forest plan to the public for review and comment, political appointees in Washington, DC, suspended the release so that they could review the draft plan to make sure it did not run counter to the Bush administration's oil and gas exploration initiative on public lands. After several weeks of review, the appointees allowed the draft plan to be released with minor modifications.

Within the USFS, much of the feedback from the Core Planning Team concerned the need to strike a balance between producing substantive outcomes and emphasizing relationship-building activities. The assessments were under a tight schedule to finish, which required the Core Planning Team to allocate its time and energy strategically. When the LWG process was not leading to the assessment product outcomes that the Core Planning Team expected, team members reacted negatively against the collaborative process. Carmine was at the center of this debate and recalled that:

> *... the Core Planning Team members had high expectations early on that the LWG process would yield products that they could simply plug into the documents for which they were responsible. When this proved not to be the case, frustrations ran high and they began to question the value of all the effort going into the collaborative work.*

As a result of this feedback, the Core Planning Team developed a strategy for working on the assessments while the LWG process continued – a 'phase shift' in the LWG process. Instead of scheduling a continuous slate of meetings spread out over six to eight months for each LWG, the LWG process was divided into two phases. The first phase was composed of two introductory meetings where participants worked together and with USFS staff to generate a vision statement, desired conditions and goals concerning the future management of the geographic area utilizing the aforementioned landscape theme activity. After these two meetings, the Core Planning Team spent time developing a draft assessment based on the vision statement and issues generated by the landscape management theme exercise.

The second phase occurred after the draft assessment was completed. The collaboration team and formal LWG process were discontinued prior to this phase due to lack of funding; however, the Core Planning Team and community stakeholders maintained the dialogue that stemmed from the LWGs. Stakeholders were convened in periodic community meetings throughout the GMUG to further explore and refine desired conditions and management options through a continuous feedback process between the Core Planning Team and community stakeholders. Between spring 2003 and summer 2006, this continuous feedback process shaped the core elements of the draft revised forest plan.

Feedback and adaptations relating to the relationship criterion

The relationship indicators focus on the degree of respectful listening and discussion, mutual learning and common ground among participants. The results showed that there was a wide variation in group dynamics in specific LWG meetings, within a geographic area and across geographic areas. One important lesson is that the collaboration team should not have over-analyzed one particular meeting or the feedback from the most dissatisfied participants, as Carmine points out:

> *Depending on the night, the group, the atmosphere or 'group dynamic', sometimes the material or activities will fall flat. It is important not to over-adjust or over-compensate. When you have materials and activities that have worked well in other settings, with other groups, don't abandon them just because you have an off night.*

Overall, the monitoring results suggested that there were new forms of dialogue among diverse stakeholders. There were consistently positive responses to group exercises where diverse stakeholders discuss issues, learn from one another and share knowledge and values for the forest. In one instance, the members of a small group activity at an initial LWG meeting insisted on staying together throughout the LWG process in their geographic area. They were truly interested in what one another had to offer about the landscape and were surprised that they could actually talk about issues in a respectful manner. Susan observed this from a community perspective:

> *'As a community we learned by respectfully listening to diverse stakeholders share their values and vision of the GMUG as we participated in formal meetings. We also learned from agency field staff and resource staff as they shared their knowledge of the forest.*

The feedback that the 'field staff', or district staff, were invaluable at the LWG meetings led to a critical adaptation. At the initial LWG meetings, district staff members were largely on the sidelines while the core planning and collaboration teams took center stage. However, district staff members have more intimate knowledge of, and experience in, the geographic areas and are members of local communities and have established relationships within these communities. They are field-level sources of information and feedback to the Core Planning Team, as well as the public face of the USFS in local communities. After six months of LWG meetings, the Core Planning Team made a strategic decision to involve district staff in planning and playing a more active role in the LWG meetings. By participating in the design and facilitation of LWG meetings, district staff enriched the dialogue among stakeholders and lent legitimacy to the LWG process.

The involvement of the district rangers and staff was paramount to the building of collaborative relationships, not only between participants and the agency staff, but within the GMUG forest staff. Their involvement enhanced the credibility of the process by serving as sources of information and members of the communities. Similarly, local government officials' support and participation in the LWG process was key in terms of lending legitimacy to the process. Delta County commissioners were advocates of the collaborative learning process from the beginning and have actively participated in the process.

The relationship monitoring feedback provided an overall picture that the LWG process was producing positive relationship-building outcomes. However, there was no indication that a great amount of trust was built through the LWG process between the USFS and participants, and especially between the USFS and non-participating stakeholders. In fact, the lack of delivered products at the promised date negatively affected trust-building because community stakeholders had done their part in participating in, and contributing to, the LWG process; but their contributions had not been reciprocated with a promised product.

LESSONS LEARNED: MONITORING AND ADAPTIVELY MANAGING A COLLABORATIVE PROCESS AS A TEAM

Each collaboration team member learned important lessons about working through the creative tension resulting from expectations about roles in, and outcomes of, the LWG collaborative process. For Carmine, the LWG process exposed fundamental challenges in institutionalizing collaboration as standard practice in the USFS:

> *My creative tension centered around the need to revise the GMUG forest plan within existing time and budget parameters, all the while improving relationships with multiple internal and external stakeholders and ensuring that a fair, open, rigorous, legally defensible process has been achieved. Collaboration is slow and messy. Taking up that much time and energy can lead to frustration on the part of everyone involved because we all have other tasks we absolutely need to get done, like writing the assessments and draft plan. The key is to keep lines of communication open.*

Collaboration with stakeholders is in addition to all of the other legal and administrative responsibilities Carmine and the Core Planning Team are obligated to perform. For Kathy, her role evolved to be more encompassing of all stakeholder interests, including Carmine, the Core Planning Team and Tony:

> *As a professional mediator/facilitator, I was ethically bound to represent all stakeholders in the LWG process. My negotiation skills were in high demand as I sorted through stakeholder positions and interests – from members of the Core Planning Team to user groups with national constituencies to an academic researcher. At issue was not only how to negotiate between stakeholders to come to agreement about objectives and outcomes, but also what role I should have played in the conflict.*

Kathy ensured that the process itself was not altered in a way that was perceived as serving the needs of only a few stakeholders. Her role expanded to not only plan, facilitate and mediate LWG meetings, but to facilitate and mediate the needs of various stakeholders outside of the LWG meetings.

For Susan, the LWG process was a reminder of the challenges that communities face as a partner with a federal agency:

> *My expectation as a partner was not to have just participated in public meetings but to have been part of a broader, more formalized, steering committee that oversaw the whole process from the community perspective. Instead, the adaptive management emerged primarily within the agency. Another issue is that agency staff did not recognize the resourcefulness of the community itself to help them accomplish the objectives of the collaborative learning process. It was a learning process for the agency to work collaboratively with a community.*

The LWG process never fulfilled its promise of being a 'community-based collaboration' because the community was not involved in shaping its design, monitoring and adaptive management. However, the feedback of Susan and other community leaders has been vital for calibrating the collaborative process.

For Tony, the creative tension has clarified some critical distinctions in researching collaborative processes:

> *The process monitoring and adaptive management approach to the LWG effort was my first real participatory action research endeavor. It involved a lot of negotiation over what questions to ask and what tools were appropriate. It also entailed a lot more critical questioning of my role and research results, which wasn't always comfortable and produced some interpersonal conflict. I came to appreciate the rigors of partici-*

patory action research. Participatory research models collaboration – by its nature it is collaboration.

Academic researchers, especially social scientists in natural resources, have taken a great interest in collaboration in natural resources. They are trained to conduct research at an arm's length with valid, replicable research results, for these are the projects that can be published in peer-review journals. These professional norms can run counter to the research needs of a collaborative process. Tony's comments suggest that more attention needs to be given to providing training, mentoring and rewards for academic researchers partnering in participatory action research in collaborative processes.

The monitoring and adaptive management of the LWG process developed in an *ad hoc* fashion since it grew out of a need to continuously find the best fit between the substantive requirements of the GMUG forest plan, stakeholders' sophistication and expectations, and the collaboration and Core Planning Team's capacity to carry out a collaborative learning process. This evolution was underscored by a creative tension resulting from the collaboration and Core Planning Team's assumptions and expectations. Below are recommendations for the growing number of collaboration efforts in public forest planning.

Build in a systematic monitoring approach of the collaborative process from the start, rather than react to crises

The need to learn and adapt from each collaborative interaction is a given. A multi-party, participatory design, monitoring and adaptation approach is highly effective in maximizing learning. Key individuals on the 'collaboration team' include agency planning and field staff, community leaders, the third-party neutral facilitator and researchers who are interested in conducting evaluation research on the process. It is essential for the collaboration team to take time in the early phases to identify criteria, indicators and monitoring mechanisms that provide multiple sources of feedback and information about the collaborative process. One suggestion from the LWG process is to establish a community group to provide perspectives on how to adaptively manage the collaborative process to effectively address stakeholder issues and concerns. This was not done for the LWG process because of the lack of agreement of who would be responsible for convening such a committee. Additionally, the USFS is constrained by the Federal Advisory Committee Act in establishing steering committees without a formalized process, so some care should be given as to how this group is organized, as well as to its scope of work.

Clearly define and test the appropriateness of criteria and indicators of progress

Identifying indicators in advance is a challenge because there are many unknowns about collaboration before it actually happens. Treating the process monitoring and adaptive management approach as an experiment can be risky because people are dedicating substantial time and energy to these processes. Adjusting feasible criteria and indicators will contribute to the calibration of the collaborative process. The indicators must target specific measurable components for each criterion and will likely be both qualitative and quantitative. Due to the subjective nature of qualitative indicators, multiple monitoring mechanisms to triangulate feedback were used for the LWG process. This triangulation approach provided a richer picture of the collaborative process than relying on one feedback mechanism.

Clearly define roles and responsibilities in the collaborative process and the monitoring component

It is essential to clarify who is responsible for collecting, analyzing and interpreting feedback information and how the information is compiled and distributed to the rest of the team, who are part of the adaptive management process. An important part of this clarification process is explicitly acknowledging assumptions and expectations that each team member has about their role in the collaboration and about how they view the adaptive management of the collaborative process. These have to be open, honest and frank discussions because each team member generally expects a certain outcome from the process and in an attempt to attain it, may jeopardize the fairness, openness and, therefore, legitimacy of the process. It is better to have these discussions prior to the collaborative process, rather than half way into it, where stakeholders are already engaged and expectations are high.

Retain flexibility in the monitoring mechanisms to fit the situation

In order for monitoring mechanisms to produce meaningful and useful feedback, they must be tailored to fit the situation. Formal mechanisms such as questionnaires, regularly scheduled interviews and debriefing meetings can set the stage. But it is often the informal mechanisms – spontaneous phone calls or chance meetings on the street or in the grocery store – that can generate the most meaningful, useful information about the collaborative process. People tend to communicate their knowledge best in specific situations – a knowledge-in-action mode. It is critical to identify where these opportunities may exist and to capitalize on the opportunities.

Frequently communicate with all stakeholders – agency staff, community members and organized groups – to ensure that the collaborative process is flexible

The collaboration team was upfront with all participants in the LWG process that the process was being actively monitored and adaptively managed to keep up with evolving expectations, issues and information. In doing so, the participants expected change, even if some of the changes were difficult to understand initially. Participants also expected to contribute to the monitoring process by filling out questionnaires, having phone or personal interviews or contributing in meetings.

Frequently communicate with all stakeholders about what and why adaptations are being made to the process

Participants deserve to be informed about the status of the process and not be surprised when things suddenly look and feel different. When they are informed, they are more likely to support and contribute to the collaborative process. A steering or advisory committee to help guide the collaborative process can be a great benefit because committee members can voice concerns that other participants have about the process, as well as help to inform participants about process changes. It is also important to be clear about scheduled timeframes for deliverables so that stakeholders know when to expect a tangible product to work with. If there are changes to the scheduled timeframe, stakeholders need to be informed of what changes are made, why they are made and need to be given a new scheduled timeframe.

Final thoughts

Collaboration, like ecosystem management, is essentially an experiment in getting people with different values, experiences and knowledge to work together to achieve goals that they could not achieve alone. The implementation of a collaboration process feature or technique provides an opportunity for learning how it worked relative to the substantive issues, openness and fairness of the process and working relationships. Continuously monitoring, evaluating and adapting the collaborative process can help to generate progress in all three areas. This chapter exposed the details of a collective effort to monitor and adaptively manage a collaborative process. Not only were specific indicators and measurement techniques described, but so were how the assumptions and expectations of individuals involved in the team charged with managing the collaborative process contributed to a creative tension in defining indicators, interpreting feedback and designing process adaptations.

Perhaps the most enduring lesson is how the collaboration team modeled and mirrored the LWG collaboration by emphasizing learning and constant dialogue – 'collaboration within collaboration'. This was a new experience for each team member, each of whom was used to working in his or her own professional niche with well-defined roles, responsibilities and expectations. The LWG process was certainly not something any one person could have designed and implemented on his or her own. It was truly a collaborative effort that evolved and adapted over time.

On a final note, this chapter offers a template for presenting 'lessons learned' from a collaborative process. Traditionally, such presentations are made from one perspective and in one voice, typically from the academic researcher (Wondolleck and Yaffee, 2000) or facilitator/mediator perspectives (Bingham, 1986). While these perspectives are important and are read by peers who seek to learn more about collaboration, they may reach only a part of the audience who can benefit from learning about collaboration. Agency planners and community leaders typically do not publish their analyses of collaborative processes. Yet, their perspectives and voices enrich the lessons from a collaborative process by offering complementary and divergent viewpoints from academics and mediators/facilitators.

Notes

1 The GMUG Core Planning Team included individuals trained in a hydrology and watershed management, wildlife biology, forest and rangeland ecology and management, recreation management, the human dimensions of natural resources, and geographic information system specialists.

2 Available online at www.fs.fed.us/r2/gmug/policy/plan_rev/proposed/Plan_4_web/!GMUG_plan_draft_March2007hrs.pdf. The work on the GMUG forest plan revision was suspended in April 2007 due to a nationwide court injunction imposed on the 2005 national forest planning rule. As of this writing, there is no clear direction for the GMUG draft revised plan.

References

Bingham, G. (1986) *Resolving Environmental Disputes: A Decade of Experience*, Conservation Foundation, Washington DC

Blatner, K. A., Carroll, M. S., Daniels, S. E. and Walker, G. B. (2001) 'Evaluating the application of collaborative learning to the Wenatchee fire recovery planning effort', *Environmental Impact Assessment Review*, vol 21, pp241–270

Bouwen, R. and Taillieu, T. (2004) 'Multi-party collaboration as social learning for interdependence: Developing relational knowing for sustainable natural resource management', *Journal of Community and Applied Social Psychology*, vol 14, no 3, pp137–153

Cheng, A. S. and Fiero, J. D. (2005) 'Collaborative learning and the public's stewardship of its forests', in J. Gastil and P. Levine (eds) *The Deliberative Democracy Handbook: Strategies for Effective Civic Engagement in the Twenty-first Century*, Jossey-Bass, San Francisco, CA

Daniels, S. E. and Walker, G. B. (1996) 'Collaborative learning: Improving public deliberation in ecosystem-based management', *Environmental Impact Assessment Review*, vol 16, pp71–102

Daniels, S. E. and Walker, G. B. (2001) *Working through Environmental Conflict: The Collaborative Learning Approach*, Praeger, Westport, CT

Gray, B. (1989) *Collaborating: Finding Common Ground for Multiparty Problems*, Jossey-Bass, San Francisco, CA

Hemmati, M. (2002) *Multi-Stakeholder Processes for Governance and Sustainability: Beyond Deadlock and Conflict*, Earthscan, London

Holling, C. S. (1978) *Adaptive Environmental Assessment and Management*, John Wiley and Sons, New York, NY

Margoluis, R. and Salafsky, N. (1998) *Measures of Success: Designing, Managing and Monitoring Conservation and Development Projects*, Island Press, Washington, DC

Schusler, T. M., Decker, D. J. and Pfeffer, M. J. (2003) 'Social learning for collaborative natural resource management', *Society and Natural Resources,* vol 15, pp309–326

Walter, C. J. (1986) *Adaptive Management of Renewable Resources*, Macmillan, New York, NY

Wondolleck, J. M. and Yaffee, S. L. (2000) *Making Collaboration Work: Lessons from Innovation in Natural Resource Management*, Island Press, Covelo, CA and Washington DC

Wright, P. A. and Colby, J. L. (2002) *Monitoring for Forest Management Unit Scale Sustainability: The Local Unit Criteria and Indicators Development (LUCID) Test,* USDA Forest Service, Fort Collins, CO

8

Inclusion and Exclusion: Immigrant Forest Workers and Participation in Natural Resource Management

Heidi L. Ballard and Brinda Sarathy

Introduction

Over the past two decades, federal agencies, natural resource-dependent communities and many environmental groups in the western US have sought out more collaborative, participatory and community-based forms of forestry. This approach to forestry calls for a devolution in resource management as decision-making moves from the strict purview of the federal government to include local-level communities and other groups concerned about the management of public lands. The underlying assumption is that people who depend upon forests for a livelihood have both an inherent interest in, and are key to, the sustainable management of our natural resources. However, not all groups are readily recognized as legitimate stakeholders in community forestry collaborations (McLain and Jones, 1997; Baker and Kusel, 2003). Immigrant and minority forest workers, in particular, have largely been excluded from decision-making about natural resources (Brown, 2000).

Immigrants who work in natural resource industries often lie on the social margins, are frequently not informed about land management decisions, may be limited in their English language ability and travel outside their communities of residence in order to pursue work in the woods. They are simply not considered part of the communities in which they work, using the more traditional definition of the 'community' to mean a homogenous small group of people with shared norms (Agrawal and Gibson, 2001). Despite their marginal status, we argue here that it is absolutely necessary for land managers to include immigrant and minority workers for the sustainable management of forest resources. By virtue of their hands-on experience, such groups actively shape the natural terrain, have specialized ecological knowledge about forests and have a stake in managing the resources upon which their livelihoods depend. Participatory research, with its emphasis on including people who have particular knowledge of the subject being researched, may be a way to expand conceptions of community and community forestry to include these immigrant and minority forest workers.

In this chapter, we address the exclusion of immigrant workers from the science, policy and management of natural resources, as well as explore how immigrants may successfully be included in resource management decision-making through participatory research. We highlight two cases. The first documents how Latino tree planters (known as *pineros*) and their extended families in southern Oregon's Rogue Valley have been excluded from conceptions of 'community'. Such exclusion serves to keep Latinos from participating as active stakeholders in decision-making about forest management. The second case focuses on Latino brush harvesters in Washington's Olympic Peninsula and highlights how, through participatory research, immigrant harvesters successfully participated in resource management and helped to advance ecological knowledge.

COMMUNITY FORESTRY IN THE US

To examine the potential for immigrant workers' participation in forest research and management, it is important to locate them in the discussion of community-based natural resource management (CBNRM). Community-based management programs emerged several years ago in Africa and Asia in an effort to address the shortcomings of state-controlled stewardship of open-access resources (Murphree, 1993). These programs vary widely in the amount of state control, community participation and community benefit involved and have seen both successes and failures (Barrow and Murphree, 2001). Often with an eye to the lessons learned abroad, the concept of community-based management has recently gained support in a variety of North American environmental contexts, ranging from water quality to forestry (Western and Wright, 1994; McLain and Jones, 1997). In the western US, the practice of CBNRM is perhaps most evident in numerous collaborative efforts to manage federal lands through 'community-based' forestry.

Emerging during the early 1990s, the community forestry movement in the western US was perceived by many as a potential solution to environmental gridlock over the management of national forests (Kemmis, 1992; McCarthy, 2006).[1] The 'movement' itself was comprised of various efforts by rural communities to gain control of the management of federal lands for both environmental and economic gains. Much of the literature on community forestry in the US refers to collaborations – born out of environmental crisis or conflict – between community partners and the Forest Service (Pardo, 1995; Brendler and Carey, 1998). Some of the best-known collaborations include the Applegate Partnership in southern Oregon (Sturtevant and Lange, 1995, 2003; Rolle, 2002), the Watershed Research and Training Center in Hayfork, California (Danks, 2000), and, more recently, the Rural Voices for Conservation Coalition.[2] These various collaborations have been hailed for bringing disparate groups in rural communities together to discuss resource management issues, build community capacity and provide opportunities for members to work as forest stewards. Scholars have also portrayed such efforts as a foil to the top-down technocratic approach of resource management developed in the progressive era of scientific forestry and celebrate collaboration in terms of its potential for reviving democracy and civic/folk science (Weber, 2000; Borchers and Kusel, 2003).

Despite the promise of collaborative efforts in resource management, the concept of 'community' in community forestry remains elusive. Different definitions on the part of various stakeholders result in different winners and losers. There are at least three definitions of 'community' prevalent in the community forestry movement. First, some collaborations regard only full-time residents who live in proximity to forests as the primary 'community' who should derive economic and ecological benefits from resource management (Brendler and Carey, 1998; Danks, 2000). The stakeholders within such collaborations tend to include locally based timber workers and environmental activists, the majority of whom are white and US citizens.[3]

Second, and by contrast, many national environmental groups strongly oppose the decentralization of forest management to local communities. Michael McCloskey (1999, p624), former chairman of the Sierra Club, argues that 'enchantment with localism' shifts the imperative of resource management from national standards to ideals of 'home places'. McCloskey perceives these 'home places' (or local forest communities) as captured by timber and business interests. When it comes to public lands, McCloskey thus notes that all citizens, whether near or far, should have a say in the management of federal land. It is notable that this perspective neither questions what the 'majority' or 'national' interest constitutes, nor makes room for the participation of non-citizens (who might play an active role) in managing federal land.

Finally, a third and more inclusive definition of 'community' seeks to include individuals who may not necessarily be full-time local residents or even citizens, but are tied by their livelihood practices. This concept of community recognizes immigrant and migratory forest workers' knowledge of natural resources and sees their participation in resource management as valuable. McLain and Jones (1997) cite evidence to show that mushroom harvesters, who visit sites year after year, develop a sense of attachment to these places, actively steward the land and contribute economically to the local areas through their commerce. Ballard and Huntsinger (2006) and Lynch and McLain (2003) have shown similar patterns of forest stewardship by Latino floral greens harvesters in the Olympic Peninsula of Washington. These scholars argue that 'mobile workers' (mobile in terms of their work), many of whom are also non-white immigrants, need to be included in definitions of 'community'.

Excluding such groups from resource management not only affects forest workers, but may also impact upon the resources in question, as well as local economies. Moreover, it reveals how community-based forestry's socio-economic aspects have potentially become a narrow debate about local place-based workers, rural entrepreneurship and niche markets, rather than one which grapples with larger structural issues posed by the part-time/seasonal nature of natural resources work, the vulnerability of low-income and immigrant forest workers and the low-wage, no-rights situation for all contingent workers in the US (Brown and Marín-Hernández, 2000). As Brown and Marín-Hernández (2000) acknowledge, there will always be a battle for 'legitimacy' in community forestry: which workers count and why. While this struggle will continue, they argue that the important issue for community forestry is to include all workers in its analysis and efforts.

While these varying definitions of 'community' highlight some of the exclusionary tendencies within community forestry collaborations, they do not show how or why certain groups, particularly non-white and immigrant forest workers, may be marginal to science and decision-making about natural resource management. We present here two cases, respectively, that shed light on:

1 various barriers to Latino forest workers' participation in community politics and natural resource management; and, conversely,
2 how such workers may be included as active participants in natural resource management and contribute to local ecological knowledge.

CASE STUDIES

Pineros in the Rogue Valley, Oregon

Most Mexican immigrants initially came to Oregon on the *bracero* program during the early 1940s (Massey et al, 2002). The Bracero Accord, established in 1942, was a temporary worker arrangement between Mexico and the US (Gamboa, 2000). This arrangement, which lasted over 20 years, allowed over 4 million Mexicans to work legally in the US (Massey et al, 1987). The Bracero Accord ended in 1964. Other guest worker programs have since followed. During the *bracero* period, workers generally returned to Mexico once agricultural harvests were over. By the late 1970s and early 1980s, however, single male workers began to settle in southern Oregon to pursue opportunities for tree planting on federal lands in the region (Hartzell, 1987; Mackie, 1994).

Latino immigrants were recruited to plant trees on federal and private lands in southern Oregon's Rogue Valley, first by Anglo forest labor contractors and later through their own networks of kin. By the mid 1980s, some Latino forest workers (otherwise referred to as *pineros*) had assumed supervisory positions on tree-planting crews. During the late 1980s and early 1990s, a number of factors led to the gradual 'Latinization' of the forest workforce in Oregon. The 1986 Immigration and Reform Control Act, for instance, gave amnesty to numerous undocumented immigrants and opened up opportunities in labor contracting for *pineros*. Newly established immigrant contractors were able to take advantage of their social networks to recruit other Latinos (often undocumented) into forest work. By the early 1990s, significant numbers of Latino immigrants began to settle in the Rogue Valley, partially as a result of opportunities to do manually intensive forest management work (Sarathy, 2006).

Pineros constitute a non-unionized and seasonal workforce that faces harsh labor conditions, with little recourse to workplace violations such as untimely pay, insufficient pay for hours worked and a lack of safety equipment (Knudson and Amezcua, 2005). On federal lands, the Forest Service and Bureau of Land Management contract out ecosystem management work to immigrant crews. Ecosystem management activities include labor-intensive operations such as tree planting, cutting small diameter trees, pesticide application, gopher baiting, thinning brush and generally reducing heavy fuel loads in the forest understory (Beltram and Evans, 2001).

Despite their dominance and experience in forest work, *pineros* in the Rogue Valley neither participate in decision-making nor inactivism about the management of federal lands. Ironically, however, the Rogue Valley and southern Oregon, more generally, are renowned for environmental activism. In the summer of 2002, the Biscuit fire complex scorched over 202,000ha (500,000 acres) of federal and private forests in the region and led to heated debates and public actions about whether to salvage log fire-killed trees and replant burned areas. The majority of participants at the almost monthly rallies in front of the Forest Service Headquarters in downtown Medford, Oregon, were white environmental activists and their opponents, white loggers. However, members of the valley's Latino population – many of whom work in the woods and would also be directly affected by decisions regarding forest management – were significantly absent from such rallies.

Latinos' absence may be largely attributed to their general disconnect from the area's primarily Anglo environmental and logging groups. Latinos' absence from public rallies around forest management not only manifests immigrants' disconnect from environmental and logging groups, but it also highlights a lack of outreach to immigrant communities on the part of activist groups and organizations within the Anglo community. In addition, the dearth of national and local media coverage – which tends to focus more on environmentalists and loggers than on immigrant forest workers – exacerbates the erasure of Latinos from the place and environmental politics of the Rogue Valley (Sarathy, 2008).

Pineros' exclusion from community politics and natural resource management

How can we understand such instances of immigrants' non-participation and socio-political exclusion from resource management, particularly in the context of a community forestry movement that may champion the 'local community' but not immigrant forest workers? Certainly, language barriers can deter immigrants from participating in public events and rallies, as well as outreach on the part of government agencies. For example, the US Forest Service Headquarters in Medford primarily circulates information on land management plans and holds public hearings in English. Without access to Spanish language interpretation and active outreach to immigrant communities, most *pineros* are not informed of resource management decisions and remain ignorant about how they may participate in such decision-making.

The undocumented legal status of many immigrant forest workers is another factor hindering Latino immigrants' participation in natural resource management. When interviewed, many *pineros* openly admitted that they did not want to 'rock the boat' or draw attention to themselves for fear of losing their jobs or being deported. Many forest workers also do not trust government officials and for justified reasons. At least one incident, in which immigrant forest workers were arrested and deported by Immigration and Customs Enforcement (ICE) authorities, stemmed from communication between ICE officials and the law enforcement branch of the US Forest Service.[4] In general, the heightened context of raids on immigrant workplaces, including those in forests of the Pacific Northwest, have served to heighten fear and confusion within the immigrant community and to marginalize workers even further (Jenkins, 2007).[5]

Yet another factor inhibiting *pineros'* participation in natural resource management and activism more generally is the issue of intra-ethnic exploitation and hierarchical labor relations. Put simply, most *pineros* are recruited onto labor crews through kinship and social networks. As such, forest workers' immediate employers (forest labor contractors) and supervisors (crew foremen) also tend to be relatives. It is not uncommon, for example, to see entire crews composed of members from the same extended family, with labor contractors and crew foremen hiring their brothers, cousins and nephews. What is significant here is that labor contractors and foremen most often have legal status (either US citizenship or permanent residence), while a significant number of their employees are undocumented. Such structural inequalities in legal status between employers and workers, compounded by *pineros'* feelings of obligation to their kin/employers, reveals why many immigrant forest workers may be reluctant to get involved in labor-related activism or even speak out about natural resource management more broadly.

Other structural factors also hinder the participation of immigrant forest workers from natural resource management collaborations and decision-making. Indeed, federal land management agencies such as the US Forest Service and

Bureau of Land Management usually do not provide incentives nor have an interest in soliciting the experience-based ecological knowledge or opinions of forest workers. Rather, land managers are primarily interested in the effective and timely execution of ecosystem management services: workers are contracted to get the job done, not provide input on the ecological merits or consequences of various treatments.

Finally, it is important to recognize that historical legacies of marginalization may also influence Latino immigrants' decision to *not* participate in contemporary resource management politics. The following section briefly highlights some of the historical experiences of social hostility and exclusion as narrated by *pineros* and their family members who first settled in the Rogue Valley.

Historical experiences of social exclusion

Numerous Latinos and Latinas related stories of marginalization when asked about their early years in the Rogue Valley. The first Latino/as to settle in the area during the early 1970s faced significant physical segregation and social hostility. Alejandra Puentes recalls a particularly public instance of such hostility:

> *I remember one time we had just gotten here and we had an old 67 Mustang. And I remember my dad had it, like, jacked up in the back. And the family was together and we were driving down Riverside. And there was a little old man who wanted to cross the street. So my dad stopped and let him cross. And when he got to the island in the middle of the street, he stopped. I'll never forget this – it was an impact, as we had just gotten here, you know. He stopped in front of our car and with his cane, he started showing it to us and said: 'Go back where you came from. Leave us alone, you're not wanted here.'*

The humiliation experienced by the Puentes family is particularly harsh due to the public manner in which they were insulted and graphically illustrates the ways in which Latinos are not considered members of the community by traditional definitions. Alejandra's father was verbally shunned from the central street of downtown Medford and metaphorically cast out of the city's core. As new arrivals seeking to make the Rogue Valley home, they were told that their highly visible presence in the heart of the city was unwelcome. By threatening the Puentes family both verbally and physically, the old man essentially engaged in policing the boundaries of where Latinos could or could not be.

Like the old man, many Anglo residents preferred not to interact with Latinos in the Rogue Valley, who were regarded with fear and social anxiety. The following speaker notes that although Latinos' presence in the region was never publicly acknowledged, 'everyone' felt relieved when they left:

> *Medford used to be a lily-white community for so long. It's only natural for people to feel scared about newcomers. They were different... There would be thousands of workers coming into this area for picking pears. No one ever talked about them. Everyone just breathed a sigh of relief when the season was over ... I used to be so scared when my husband was out of town. We used to live next to the pear orchards and when he was gone I wouldn't be able to sleep the whole night.*

Indeed, Latino/as were relegated to and tolerated only within certain cordoned off spaces, such as fields and orchards on the margins of town. Immigrants were also expected to leave after the fruit season was over. In driving through downtown Medford and *settling* in the area with his family, Alejandra's father thus breached many Anglo residents' unvoiced expectations about Latinos' mobility: to remain invisible, on the peripheries of town, and to reside in the area only temporarily.

The case of *pineros* (and the larger Latino community) in the Rogue Valley reveals some of the challenges that immigrants face when it comes to participating in the community and public decision-making more generally. We have seen how language barriers, insufficient outreach on the part of federal land management agencies, undocumented status and intra-ethnic exploitation can contribute to many Latino forest workers' exclusion from natural resource management. In addition, a legacy of social exclusion and hostility only reinforces the non-participation of minority groups in the management of natural resources and other issues that concern their lives and livelihoods.

So how can these minority groups be included in community forestry if they are not even included in the community by traditional definitions? In fact, it is precisely their participation in natural resource work that provides a bridge to both the community and community forestry. If we choose to follow the more inclusive definition of community forestry that includes all those who live and work in the forests, how might collaborative efforts at resource management be more comprehensive? Latino forest workers possess two crucial characteristics that make them essential and important potential contributors to forest research and management despite the cultural, linguistic and often legal barriers: livelihood dependence upon the forest resources and local ecological knowledge of the forest. The following case of floral greens harvesters in the Olympic Peninsula, WA, illustrates an instance in which Latino forest workers actively participated in decisions about natural resource management and how participatory research helped to enable this process.

Floral greens harvesters in the Olympic Peninsula, Washington

There are a number of ways that a group of people may participate in the management of natural resources in the US. These actions range from attending group meetings about resource management, negotiating with the Forest Service, writing letters to locally elected officials about resource management concerns, formulating bill concepts on resource management practices, lobbying for resource management policies, engaging with other organizations on these issues, and directly managing the natural resources on the ground. Most of these activities are not accessible to the majority of immigrant forest workers, save the last. *Pineros* and other forest workers may have decision-making power on a small scale with regard to managing the forest and understory on the ground (e.g. where a tree is planted and what plant is harvested), but almost no influence on a large scale (e.g. how many acres are planted and the duration of time between harvests). However, one way in which immigrant forest workers *can* participate in natural resource management on a larger scale is by participating in the science upon which the management is based.

In the Pacific Northwest, many communities, land managers and scientists are turning to non-timber forest products (NTFPs) as an alternative or complement to timber production. In addition to mushrooms, among the most lucrative of these NTFPs are the floral greens used in decorative floral arrangements. These greens are generally understory shrub species – such as evergreen huckleberry (*Vaccinium ovatum*), sword fern (*Polystichum munitum*) and salal (*Gaultheria shallon*) – that grow naturally in managed or unmanaged forests. Those who harvest floral greens, predominantly from the lowest socio-economic levels of US society, are ethnically diverse, generally have limited educational backgrounds and often speak very little English (Schlosser et al, 1991; Von Hagan and Fight, 1999). They are also generally excluded from processes that would give them a voice in the development of management approaches for public and private lands.

During the early 20th century when the industry first emerged, harvesters were primarily Anglo Americans who needed a supplementary income or simply wanted to work in the woods. The industry expanded during the 1970s when a large influx of labor in the form of refugees from war in Southeast Asia began coming from outside the US (Hansis, 2002). Then, during the late 1980s and 1990s, immigrants from Mexico and other Latin American countries discovered floral greens harvest as an alternative to agricultural work in California, Oregon and Washington's eastern fruit orchards (Hansis, 2002). These influxes have resulted in current demographics, where the majority of floral greens harvesters are from Latin America, with a smaller proportion of Southeast Asians and Anglo Americans making up the workforce.

In western Washington and Oregon, floral greens such as salal grow in abundance across a variety of landownership types: public federal and state lands, as well as private lands, including small, non-industrial timber lands and large, industrial timber company lands. However, because of the overwhelming focus on timber management in these forests, the floral greens industry has come to rely on landless harvesters who gain access to the land via permits or leases for the rights to hand pick one or more floral greens species. With little regulation within the industry, each land manager, public or private, has a different policy for selling and enforcing their permits. The rules of access can vary widely with only a few steps through the forest, where public and private lands form a mosaic across the landscape. The permit systems of neighbors can differ dramatically in duration of access allowed, cost, documentation and insurance required, and limitations on quantity of product allowed.

Despite this variety of rules and systems of access, de facto open access conditions actually prevail for floral greens in most regions in the Pacific Northwest. In the face of increasing resource demand, as the Olympic Peninsula of Washington is facing for floral greens species, these open-access tenure regimes are likely to foster unsustainable levels of harvest (Ostrom, 1990; Bromley, 1994). It is this management context in which harvesters make decisions about where, how much and how to harvest floral greens. While most floral greens harvesters are only able to participate in forest management decision-making on a small scale, choosing which patches of salal to harvest within a forest stand for which they have a permit, some have organized to form organizations (such as the Northwest Research and Harvester Association; see Chapter 9 in this volume) capable of negotiating with public land managers and landowners for the management of NTFPs and even some timber management practices. Gaining a seat at the table with these landowners, as well as the wholesalers who buy the floral greens for export, has meant that harvester organizations (both formal and informal) are combating the exclusionary practices that harvesters usually face.

Floral greens harvesters' ecological knowledge

The particular livelihood practices of floral greens harvesters often result in their having extensive local ecological knowledge of the forests in which they work. Hence, despite the social exclusion of immigrant and minority forest workers from both the local communities and forest management decisions in the Pacific Northwest, experienced harvesters have the potential to be important contributors to ecological research on the species they harvest. As an example – the inclusion of immigrant forest workers in natural resources science and management – we present a project involving the participation of dozens of floral greens harvesters on the Olympic Peninsula, Washington, called the Olympic Peninsula Salal Sustainability and Management Research Project (hereafter called 'the Salal Project'). Dozens of floral greens harvesters participated in the Salal Project from 2001 to 2003, which was a collaboration between the Northwest Research and Harvester Association (NRHA) and a plant ecologist (Heidi Ballard), a graduate student from University of California, Berkeley, at the time. The NRHA is an organization founded by and for harvesters in 2001 to increase access to land as well as provide monitoring, research and training for harvesters. As part of the Salal Project, participant observation and semi-structured interviews were conducted with salal harvesters to document their understanding of the ecological dynamics of the resources and to ask for their participation in an ecological research and management project. Latino, Anglo and Southeast Asian harvesters were asked about what forest conditions promote commercial salal growth and how different harvest techniques and intensities affect its growth and reproduction. Harvesters were from a variety of backgrounds, including people of different genders, ethnicities, countries of origin and experience in the industry.

Results provided insight into harvester livelihood practices and ecological knowledge that is valuable to public and private land managers, as well as ecologists. Unlike many types of seasonal work that only occur for a few months each year, floral greens harvesting occurs from July through April, so that many harvesters work almost year round in the same area. They most often return each year to the same towns and areas of the forest to pick salal, developing a deep knowledge of the ecology of that particular area. In addition, because of the wide-ranging extent of the commercially productive floral greens areas, harvesters also gain a broad perspective of ecological zones and variations in habitat types. Public and private landowners and managers, on the other hand, are generally unfamiliar with the harvest, management and marketing of floral greens in these forests. This may be because the intensification of the industry has been so recent and exponential, and because only recently are some forest land managers trained multiple-use management of timber, wildlife and non-timber products.

Results suggest that many floral greens harvesters possess the motivation to sustainably manage the species upon which their livelihoods depend (Ballard and Huntsinger, 2006). Several harvesters who regularly use short-term (two-week) permits explained that they know that the way they harvest 'hurts the plant', but if they had an exclusive lease of their own for several years they would use a less intensive harvest method. With the understanding that floral greens harvesters possess both the motivation to sustainably manage the resource upon which their livelihoods depend and the ecological knowledge required to manage, monitor and even experiment with management practices within a timber production context, a question can be asked: *how can immigrant forest workers be included in current forest management, policy and research?* The Salal Project attempted to address the exclusion of harvester stakeholders from the scientific applied ecology research upon which so much forest policy and management is based. By using a PR approach, the research process and project provided concrete links between harvesters and public and private land managers, and potentially provided a firmer foothold for the consideration of civic science use within public forest management agencies.

Participatory action research

Key principles for participatory research were applied through this project in an attempt to reverse the effects of the exclusion from forest science that harvesters had experienced up to that point. These principles included building on the strengths and resources within the community (local knowledge), facilitating collaborative partnerships in all phases of the research, promoting a co-learning and empowering process that attends to social inequalities, and integrating knowledge and action for the mutual benefit of all partners (Park, 1993; Israel et al, 1998). A specific description of how these principles played out in the Salal Project can be found in Chapter 9; but here we relate how including harvesters and their local ecological knowledge resulted in better applied ecological research and benefits to the harvester community.

During the Salal Project, harvesters were involved in nearly all phases of the research, though in some phases more intensely than others. Harvesters helped to define the research question to focus on the effects of harvest intensity and helped to design the experiment (different levels of harvest intensity and where to locate plots) and specific variables on the plant to be measured (number of new shoots per square meter rather than only percent cover). Harvesters helped to collect data in the field for the three summer seasons of 2001 to 2003 (data sheets were translated into Spanish) and helped to interpret the results at a gathering of the Northwest Research and Harvester Association in which graphs of harvest yield data were analyzed. At that gathering and subsequent conversations, harvesters suggested that while initially counter-intuitive (the most heavily harvested plots produced more new growth), the results of the research showed that a rest-rotation management regime for salal would likely promote sustainability (Ballard, 2004).

As a result, harvesters contributed ecological and commercial knowledge of salal that was invaluable to the research on the effects of harvest intensity on growth and production. This resulted in a project that addressed the precise questions of harvesters and land managers concerned about over-harvesting that simply could not have been conducted by ecologists alone (Ballard and Fortmann, 2007). In addition, the harvesters often participated in teams with US Forest Service technicians so that the project served as an informal showcase of harvester knowledge and management skills, and tangibly improved relations and attitudes between harvesters and public land managers in that area. Technicians remarked on the depth of harvesters' knowledge of the ecology and, particularly, timber management practices (realizing that harvesters must work within the timber management timelines and constraints). Results of the project were distributed to land managers and scientists in the region, highlighting harvesters' participation in the research. The Salal Project's interactions with land managers and dissemination of results likely helped to inform subsequent permitting policy changes for both the state and federal forest lands in the area, which, in most cases, improved systems of access for harvesters.

Discussion

Participatory research has long been applied as a tool of community development and improved natural resource management; but the particular case of immigrant and minority forest workers in the US poses a unique and challenging set of factors. Similarly, the historical and current social exclusion of these workers from the forest management and science that determines their livelihood seems to fly in the face of the seemingly inclusive community forestry movement gaining momentum in the US. The combination of these obstacles, rather than being insurmountable, are instead a compelling rationale to use participatory research approaches with immigrant workers on questions of access, control, decision-making and management of forest resources, as well as inclusion within the communities of the Pacific Northwest. While not without major challenges, this participatory research is possible and productive, and benefits both the research and the immigrant worker participants.

The application of a participatory action research approach to ecological research with immigrant and minority floral greens harvesters, though exceeding the expectations of many, reflected many of the challenges and successes common to participatory action research (Minkler and Wallerstein, 2003). It also revealed some of the particular constraints that often impede better inclusion of immigrant forest workers in natural resource management and policy. Issues of time, efficiency and trust between researchers and the community partners are undoubtedly present in many research efforts; but in collaborating with immigrant workers, these issues are further compounded by people's lack of legal status, language barriers and distrust of outsiders.

The legal status of immigrant forest workers plays a huge role in the ease with which researchers, managers and non-governmental organizations might promote inclusion in natural resource management. Many workers in the Pacific Northwest are *sin papeles* (without papers), and contacting workers through contractors proves difficult. In the case of *pineros*, the fact that many forest workers were related to their employers inhibited their willingness to participate in collaborations that might have jeopardized their employers, supervisors and themselves. The undocumented status of many floral greens harvesters in western Washington also created obstacles to participatory research on salal. For example, introduction to the community via federal agencies, which might have proven helpful in other circumstances, was relatively unproductive because harvesters were justifiably wary of any possible connection to the Immigration and Customs Enforcement.

While limited English proficiency also presents a challenge to participatory research collaborations between immigrant workers and researchers or managers, this hurdle may be overcome with strategic outreach and inclusive practices. An overwhelming majority of floral greens harvesters on the Olympic Peninsula and *pineros* in southwestern Oregon speak English as a second or even third language. Whether a research project is conducted in English or the workers' native language will certainly influence the time required and the validity of the project. Choosing one language or another means that large numbers of participants will be excluded; but it would also likely be prohibitive for the researcher to study the several languages of forest workers enough to become fluent. In the case of the Salal Project, the researcher worked with teams of harvesters in which at least one person spoke English, and this certainly constrained the direct participation by non-English speaking harvesters. In the case of the research with *pineros*, the researcher could communicate in Spanish. For the research to truly benefit the community in its process and product, communication barriers should be reduced as much as possible, perhaps by the use of volunteer translators in small group meetings, as well as translating all announcements and newsletters into languages spoken by workers.

In the case of the Salal Project, the graduate student researcher's inexperience and outsider status with the community (an Anglo woman coming from a large university in California) was both a help and an impediment. An outsider to the region may not be perceived to have vested interests or affiliations that could be threatening, but may also not have a compelling reason to be trusted. This increased the time required for the project in order to develop trust between the researcher and a group of participating harvesters. Similarly, the outsider status of the researcher in the *pineros* case had its advantages and disadvantages. The researcher spent considerable time meeting and working with community members, and especially immigrant youth, in various contexts. Indeed, she was able to connect with many forest workers primarily through the trusting relationships built with the spouses and children of *pineros*. At the same time, because the researcher also interviewed contractors, foremen and government officials, her access to some groups of workers (and contractors) remained limited due to lack of trust.

Participatory research with immigrant workers also presents a challenge due to the nature of the natural resource work that they do. In the case of most ecological experiments, an ecologist hires one or more field assistants who are often university students or field technicians to conduct data collection during the field season. Using a participatory research approach for the Salal Project, however, required the training of several harvesters, often with very little educational background or literacy to collect data about plant variables, including consistent measurements with a compass and clinometer. This took time. In addition, many harvester participants had to leave for weeks at a time to take advantage of temporary land access or a seasonal species that needed to be harvested. Hence, new harvesters had to be trained every few weeks to ensure continuous data collection, requiring extra care in the training for consistency in data collection. For both of these reasons, the Salal Project took longer and had a smaller sample size than a more conventional ecological research project would have. Some harvesters remained consistent throughout the project, however, and provided a core for the research team despite the attrition. Furthermore, as mentioned above, the ecological knowledge and understanding of harvest practices that harvester field assistants contributed was invaluable to the project and could certainly not have been provided by university undergraduates or even other ecologists.

In comparison, the nature of collaboration and participation in these two cases differ significantly. In the Salal Project, immigrant harvesters participated in ecologically oriented research with relatively well-defined roles. It was in their self-interest to show land managers – who controlled access to the resource – that harvesters had ecologically valuable knowledge about the sustainable management of salal. Harvesters thus had a direct incentive to participate in the Salal Project: the results had the potential to justify their ongoing access to a resource upon which their livelihoods depended. By contrast, the research with *pineros* was more ethnographic in scope and sought to document the history and experiences of Latinos in forest work over time. It was participatory in that immigrant community members actively shaped the questions, interests and agenda of large parts of the research (i.e. documenting some of the labor history of their community). However, while the research focused primarily on forest workers, it was not only *pineros* who participated in the project: friends and family of *pineros*, government officials, social service providers and many others were integral to the research. Indeed, directly connecting with forest workers was often challenging, given their long workdays and remote worksites. Researchers must thus work through involvement in community activities and events or through institutions such as churches, schools and English as a Second Language programs. To work through contacts, however, researchers must establish trust and legitimacy – often easier said than done – with both community members and institutional gatekeepers.

CONCLUSION

Natural resource management, which includes access and control of resources, is increasingly touted as relying heavily on science: scientific principles, scientific methods and scientific research. Hence, migrant and mobile forest workers who are not considered part of the local community are not only excluded from policy arenas, but also from the scientific research and knowledge production that shape resource policy, management and access. Though NTFP harvesters, for example, are not currently considered important participants in both forest management and forest research, this chapter suggests that they should be. By learning from some of the challenges and successes of these cases, scientists and managers may take a step closer to a more effective integration of academic and civil science.

A lesson from both the *pineros* and harvester cases is that organizations, no matter how informal or diverse in constituents, can play a key role in the success or failure of increasing participation by immigrant and minority forest workers in research and management. In the case of the *pineros*, relationships between the researcher and social service providers and non-profit organizations were essential for working and building trust with workers. For the Salal Project, the fortuitous collaboration with the Northwest Research and Harvester Association (NRHA) is arguably the dominant factor that facilitated harvester participation in many of the steps of the research process during the Salal Project. The NRHA has been successful in organizing harvesters from different ethnicities, language backgrounds and levels of harvest experience. This is due in no small part to their focus on their livelihood practices and issues of access to land. This focus on harvester livelihood concerns is what made the collaboration for participatory ecological research as effective as it was and involved harvesters in ways that few thought possible given their 'invisibility' as a community. The essential role of the NRHA in the participatory research on sustainability of salal harvest, therefore, may serve as an example of recent theories suggesting that institutions and their interactions, rather than traditional characteristics of presumably conservation-oriented communities (small size, homogeneity and shared norms), should be the focus of studies of community-based natural resource management and conservation (Agrawal and Gibson, 2001).

Immigrant and minority forest workers shape and maintain natural resources in their day-to-day activities, often developing extensive local ecological knowledge. Given that ecological knowledge is necessary for the sustainable management of natural resources, these workers should have access to more formal arenas of participation in natural resource management as well. This inclusion makes for a more equitable community forestry movement, as well as a more viable resource management agenda. Their participation also expands notions of who and what constitute a community, demanding the inclusion of those who work and live in the forest regardless of legal status and cultural and language barriers. It is important to note, however, that not all research projects studying issues of concern to and about immigrant forest workers necessarily lend themselves to a participatory research approach. Challenges include a lack of incentives for workers to participate, conflict between different types of forest workers and, most importantly, the safety and security of workers themselves. Participatory research is not a panacea but a tool among many others that can be used to increase the participation of immigrant workers in natural resource management.

NOTES

1 Baker and Kusel (2003) argue that this is actually the second wave of community forestry in the US. The 'first wave' concerned the Forest Service's focus on local

economic benefit through sustained yield programs, whereby rural communities were to maintain steady employment through timber harvesting.

2 See www.sustainablenorthwest.org/programs/rvcc.php.

3 Certainly, some community forestry collaborations do not include numerous working-class white workers either.

4 The incident in question involved a number of *salal* harvesters being arrested by ICE officials. The harvesters had gathered at the Mount St Helens National Volcanic Monument Headquarters in Amboy, WA, on the Gifford Pinchot National Forest to collect harvesting permits. There, ICE agents, who had been in communication with Forest Service Law Enforcement, apprehended and arrested six harvesters (Jefferson Center for Education and Research, 2005).

5 In the past year, immigrant forest workers have been deported in Forks and Shelton, WA (pers comm, Patricia Vazquez, Jefferson Center for Education and Research, 6 September 2007). Higher profile raids have also taken place in Portland, Oregon, where over 300 workers were arrested at a Del Monte fruit packing plant (Denson and Hunsberger, 2007).

References

Agrawal, A. and Gibson, C. C. (2001) *Communities and the Environment: Ethnicity, Gender, and the State in Community-based Conservation,* Rutgers University Press, New Brunswick, NJ

Baker, M. and Kusel, J. (2003) *Community Forestry in the United States,* Island Press Washington, DC

Ballard, H. (2004) *Impacts of Harvesting Salal (*Gaultheria shallon*) on the Olympic Peninsula, Washington: Harvester Knowledge, Science and Participation,* PhD thesis, University of California, Berkeley, CA

Ballard, H. and Fortmann, L. (2007) 'Collaborating experts: Integrating civil and conventional science to inform management of salal (*Gaultheria shallon*)', in K. Hanna, and D. Scott Slocombe (eds) *Integrated Resource Management,* Oxford University Press, Oxford, UK

Ballard, H. and Huntsinger, L. (2006) 'Salal harvester local ecological knowledge, harvest practices and understory management on the Olympic Peninsula, Washington', *Human Ecology*, vol 34, pp529–547

Barrow, E. and Murphree, M. (2001) 'Community conservation: From concept to practice', in D. Hulme and M. Murphree (eds) *African Wildlife and Livelihoods: The Promise and Performance of Community Conservation,* James Currey Ltd, Oxford, UK

Beltram, J. and Evans, R. (2001) *The Scope and Future Prospects – Oregon's Ecosystem Management Industry*, Ecosystem Workforce Program, University of Oregon, Eugene, OR

Borchers, J. G. and Kusel, J. (2003) 'Toward a civic science for community forestry', *Community Forestry in the United States: Learning from the Past, Crafting the Future*, in M. Baker and J. Kusel (eds) Island Press, Washington, DC

Brendler, T. and Carey, H. (1998) 'Community forestry, defined', *Journal of Forestry,* vol 96, no 3, pp21–23

Bromley, D. W. (1994) 'Economic dimensions of community-based conservation', in D. Western and R. M. Wright (eds) *Natural Connection: Perspectives in Community Based Conservation,* Island Press, Washington, DC

Brown, B. A. (1995) *In Timber Country: Working People's Stories of Environmental Conflict and Urban Flight,* Temple University Press, Philadelphia, PA

Brown, B. A. (2000) 'The multi-ethnic, nontimber forest workforce in the Pacific Northwest: Reconceiving the players in forest management', in D. Salazar and D. Alper (eds) *Sustaining the forests of the Pacific Coast: forging truces in the war in the woods,* University of British Columbia Press, Vancouver, British Columbia, Canada

Brown, B. A. and Marín-Hernández, A. (2000) *Voices from the Woods: Lives and Experiences of Non-Timber Forest Workers,* Jefferson Center for Education and Research, Portland, OR

Danks, C. (2000) 'Community forestry initiatives for the creation of sustainable rural livelihoods: A case from North America', *Unasylva,* vol 202, no 51, pp53–63

Denson, B. and Hunsberger, B. (2007) 'Immigration raid pushes Oregon into thick of fight', *The Oregonian,* 13 June

Gamboa, E. (2000) *Mexican Labor and World War II: Braceros in the Pacific Northwest,* University of Washington Press, Seattle, WA

Hansis, R. (2002) 'Case study, workers in the woods: Confronting rapid change', in E. T. Jones, R. J. McLain and J. Weigand (eds) *Nontimber Forest Products in the United States,* University Press of Kansas, Lawrence, KS

Hartzell, H. (1987) *Birth of a Cooperative: Hoedads, Inc,* Hulogos'i, Euguene, OR

Israel, B., Schultz, A., Parker, E. and Becker, A. (1998) 'Review of community-based research: Assessing partnership approaches to improve public health', *Annual Review of Public Health,* vol 19, pp173–202

Jefferson Center for Education and Research (2005) 'A troubling partnership', *Jefferson Center News,* vol 4no 2, p5

Jenkins, A. (2007) 'Rumors and panic follows immigration raids', *KUOW Program Archive,* 27 June 2007

Kemmis, D. (1992) *Community and the Politics of Place,* University of Oklahoma Press, Norman, OK

Knudson, T. and Amezcua, H. (2005) 'The *pineros*: Men of the pines', *The Sacramento Bee,* 13–15 November, www.sacbee.com/content/news/projects/pineros, Accessed 3 January 2008

Lynch, K. and McLain, R. (2003) *Access, Labor and Wild Floral Greens Management in Western Washington's Forests,* Pacific Northwest Research Station, US Forest Service, Portland, OR

Mackie, G. (1994) 'Success and failure in an American workers' cooperative movement', *Politics and Society,* vol 22, no 2, pp215–235

Massey, D., Alarcon, R., Durand, J. and González, H. (1987) *Return to Aztlan: The Social Process of International Migration from Western Mexico,* University of California Press, Berkeley, CA

Massey, D. S., Durand, J. and Malone, N. J. (2002) *Beyond Smoke and Mirrors: Mexican Immigration in an Era of Economic Integration,* Russell Sage Foundation., New York, NY

McCarthy, J. (2006) 'Neoliberalism and the politics of alternatives: Community forestry in British Columbia and the US', *Annals of the American Geographers*, vol 96, no 1, pp84–104

McCloskey, M. (1999) 'Local communities and the management of public forests', *Ecology Law Quarterly*, vol 25, no 4, pp624–629

McLain, R. (2000) *Controlling the Forest Understory: Wild Mushroom Politics in Central Oregon*, Dissertation, College of Forest Resources: University of Washington, Seattle, WA

McLain, R. J. and Jones, E. T. (1997) 'Challenging "community" definitions in sustainable natural resource management: The case of wild mushroom harvesting in the USA', *Gatekeeper Series no SA68*, International Institute for Environment and Development, London, UK

Minkler, M. and Wallerstein, N. (2003) *Community-based Participatory Research for Health*, Jossey-Bass Inc, San Francisco, CA

Murphree, M. W. (1993) 'Communities as resource management institutions', *Gatekeeper Series no 36*, International Institute for Environment and Development, London, UK

Ostrom, E. (1990) *Governing the Commons: The Evolution of Institutions for Collective Action*, Cambridge University Press, Cambridge, MA

Pardo, R. (1995) 'Community forestry comes of age', *Journal of Forestry*, November, pp20–24

Park, P. (1993) 'What is participatory research? A theoretical and methodological perspective', in P. Park, M. Brydon-Miller, B. Hall and T. Jackson (eds) *Voices of Change: Participatory Research in the United States and Canada*, Bergin and Garvey, Westport, CT

Rolle, S. (2002) *Measures of Progress for Collaboration: Case Study of the Applegate Partnership*, United States Department of Agriculture Forest Service, Pacific Northwest Research Station, Portland, OR

Sarathy, B. (2006) 'The Latinization of forest management work in southern Oregon: A case from the Rogue Valley', *Journal of Forestry*, October/ November, pp359–365

Sarathy, B. (2008) 'The marginalization of Pineros in the Pacific Northwest', *Society and Natural Resources*, vol 21, no 8, Schlosser, W., Blatner, K. and Chapman, R. (1991) 'Economic and marketing implications of special forest products harvest in the coastal Pacific Northwest', *Western Journal of Applied Forestry*, vol 6, no 3, pp 67–72

Sturtevant, V. and Lange, J. (1995) *Applegate Partnership Case Study: Group Dynamics and Community Context*, Southern Oregon University, Ashland, OR

Sturtevant, V. and Lange, J. (2003) 'From them to us: The Applegate Partnership', in J. Kusel and E. Adler (eds) *Forest Communities, Community Forests: Struggles and Successes in Rebuilding Communities and Forests*, Rowman and Littlefield, Lanham, MD

Von Hagen, B. and Fight, R. D. (1999) *Opportunities for Conservation-based Development of Nontimber Forest Products in the Pacific Northwest*, General Technical Report PNW-GTR-473, US Department of Agriculture, Forest Service, Pacific Northwest Research Station, Portland, OR

Weber, E. (2000) 'A new vanguard for the environment: Grass-roots ecosystem management as a new environmental movement', *Society and Natural Resources*, vol 13, pp237–259

Western, D. and Wright, M. (1994) *Natural Connections: Perspectives in Community-based Conservation*, Island Press, Washington, DC

9

Comparing Participatory Ecological Research in Two Contexts: An Immigrant Community and a Native American Community on Olympic Peninsula, Washington

Heidi L. Ballard, Joyce A. Trettevick and Don Collins

Introduction

> *Science, well, we say that it is more advanced. But when it comes down to the practice ... every peasant/farmer has the practice; but with the theory, well, that we don't have because we don't have the necessary resources, we don't know how to study the plant, we don't know how; but once combining both things then it becomes stronger.* (Juan, salal harvester and researcher)

Whereas community-based natural resource management (CBNRM) (Murphree, 1993; Barrows and Murphree, 2001) and participatory approaches to economic development (Chambers, 1994) have gained significant footholds in many parts of the developing world, their introduction and integration within natural resource policy and management in the US have been more recent and tentative (Western and Wright, 1994; McLain and Jones, 1997). The particular contexts, institutions and communities in the US demand a re-examination of the methods and approaches that worked effectively in other countries. In Africa, Asia and Latin America, CBNRM projects have used participatory research approaches to studying everything from wildlife populations to crop seed variety differences (Ashby and Sperling, 1995; Taylor, 2001). However, while elements of a participatory research approach have been utilized in the last two decades in the US to improve public participation in public lands' management decisions, to conduct community needs assessments and to study community resiliency and dependence upon natural resources (Kruger and Sturtevant, 2003), rarely has it been used in ecological research on natural resource extraction. Furthermore, although a participatory approach to research has been applied in the US in fields as varied as public health, sociology, anthropology, education and economic development (Park, 1993; Chambers, 1994; Minkler and Wallerstein, 2003), 'despite the extensive body of literature on partnership approaches to research, more in-depth, multiple case study evaluations of the context and process (as well as outcomes) of community-based research endeavors are needed' (Israel et al, 1998, p194). A gap still exists in the literature for a critical analysis of cases of participatory ecological research with communities on natural resource management.

In the interest of addressing this gap, we present two cases in which a participatory research approach was used to study the harvest impacts of a non-timber forest product species locally called salal (*Gaultheria shallon*) on the Olympic Peninsula of Washington. Salal is used as floral greenery in flower arrangements and is exported around the world. The first case is the Salal Sustainability and Management Project, which centers around floral greens harvesters in Mason County, Washington, the center of the floral greens industry in the US. The majority of these harvesters are immigrant workers from Latin America or Southeast Asia. The second case is the Makah Community-Based Forestry Initiative (MCBFI) on the Makah Indian Reservation on the Olympic Peninsula. MCBFI is a project within the forestry program of the Makah Indian Tribe. The Makah have harvested salal and many other species for thousands of years for cultural and subsistence purposes, but only recently decided to explore the plant's commercial potential as part of economic development efforts for the tribe.

This analysis occurs at two levels. First, we use a framework derived from several widely accepted principles of participatory research set out by Israel et al (1998) as criteria to examine the context and process of these research efforts. Israel and her colleagues suggest that these principles 'could be operationalized and used as criteria for examining the extent to which these dimensions were present in a given project' (Israel et al, 1998, p194). Second, we address the main goals of participatory research by examining the impacts of these two participatory research attempts on, first, the community; second, the research; and, third, the management and policy surrounding the resource itself. In brief, the findings confirm some, and contradict other, assumptions about participatory research in relation to natural resource management. The findings suggest that many characteristics of communities presumed to be important precursors for participatory work are neither essential nor exclusive for successful PR. Furthermore, the aspects of the projects that most effectively achieved the goals of participatory research were those that most closely involved the specific members of the community who had thorough experience with the harvesting and management of the resource.

Participatory research on non-timber forest resources (NTFRs)

Non-timber forest resources (NTFRs), which include shrubs, moss, fungi, understory plants and parts of trees harvested for cultural, subsistence or commercial use, have been studied and promoted as a tool for conservation and sustainable development in developing countries for decades. More recently, NTFRs have received similar attention in the US as part of an emphasis on ecosystem management (Kohm and Franklin, 1998) and rural community development (McLain and Jones, 2002). However, very little research and monitoring on the ecological impacts of extracting NTFRs has been conducted in the US. Because very little is known about the sustainable harvest and management of NTFRs, a participatory research approach involving the people who know the ecology of the plant and its management through daily interaction was examined as a strategy to improve research on, and resource management of, NTFRs.

We began with three propositions in attempting participatory research on NTFRs.

First, the community will be empowered. Participation in research activities by people who are often ignored in policy and management decisions can empower communities to gain more control over their livelihoods and resources (Slocum 1995), moving towards a more equitable and just society. In addition, participatory research with NTFR users can offer examples of how civic science and conventional science can complement each other on the ground.

Second, the research will be improved. Harvesters of NTFRs possess both the incentive to study the most sustainable methods of harvest for their livelihoods and the knowledge of the forest, in general, and salal, in particular, to contribute to the ecological research process, providing information, methods and analytical insights that ecologists alone could not provide (Bradbury and Reason 2003).

Third, the management and policy related to the resource itself will be improved. As land-use policies and land managers are increasingly expected to include local communities in decision-making, a process is needed by which scientists and local resource users can form partnerships in scientific research to inform that management (McLain and Lee, 1996; Getz et al, 1999). Using a participatory approach in the process of knowledge production will create natural resource policies and management practices that can benefit both the communities and the sustainability of the resource itself (Chambers, 1994; Little, 1994; FAO, 2003).

We used several widely accepted principles of participatory research as criteria to analyze the two case studies that follow (Park, 1993; Israel et al, 1998). In an extensive synthesis of community-based and participatory research literature, Israel et al (1998) suggest that community-based research:

- recognizes community as a unit of identity;
- builds on strengths and resources within the community;
- facilitates collaborative partnerships in all phases of the research;
- integrates knowledge and action for the mutual benefit of all partners;
- promotes a co-learning and empowering process that attends to social inequalities;
- involves a cyclical and iterative process of trust-building;
- disseminates findings and knowledge gained among all partners.

These principles, mediated not only by the community context, but also by the particular characteristics of ecological research, provide the framework for analyzing the successes and limitations of the participatory approach to the ecological research conducted with both the floral greens harvesters of the Salal Management and Sustainability Research project and the participants of the Makah Community-based Forestry Initiative project.

BACKGROUND AND STUDY AREAS

The forests of the Pacific Northwest are rich with NTFRs that have been used by humans for thousands of years (Jones et al, 2002). Native American tribes on the Olympic Peninsula continue to harvest hundreds of species for food, clothing, tools, medicine and art. Only in more recent history have many products also become commercially valuable, particularly edible mushrooms, edible berries, medicinal plants and the shrubs and ferns used as greenery in floral bouquets (Savage, 1995). In fact, NTFRs have become a multi-million dollar industry in the states of Washington, Oregon and parts of northern California (Von Hagen and Fight, 1999).

Among the most lucrative of these special forest products are the floral greens, which are understory shrub and fern species that grow naturally in managed and unmanaged forests. In 1989, the floral greens industry in Oregon, Washington and northern California at the point of first wholesale transaction was valued at over US$128 million and employed or bought raw materials from 10,300 people (Schlosser et al, 1991). More recent data are not available; but estimates suggest a dramatic increase since 1989. In 1994, the state of Washington exported 80 per cent of the floral greens harvested to primarily German and Dutch wholesalers (Savage, 1995).

In addition to their economic value, NTFR species in the Pacific Northwest forest ecosystems are ecologically important for the habitat and nutrients that they supply to a variety of plant and animal species (Molina et al, 1998). The potential for joint production of NTFRs with timber can provide economic incentives for forest managers to adopt management strategies, such as longer harvest rotations, that conserve biodiversity and other ecological values (Oliver,1994; Alexander et al, 2002, Kerns et al, 2003). Although previous ecological research has been conducted on salal's response to timber management practices, such as thinning and fertilizing (Bunnell, 1990; He and Barclay, 2000), no scientific literature is published on the effects of commercial harvesting of salal itself.

In a mosaic of landownership types, including public and private lands, small and large industrial timber landowners, and tribal lands, the systems of access to floral greens are as complex as the mosaic of ownership. People who harvest commercial floral greens are predominantly landless harvesters who must buy permits or leases to gain access to the plants. With little enforced regulation within the industry, each land manager, public or private, has a different policy for selling and enforcing their permits. Federal and state public land managers, who have historically focused on timber management, are now managing wildlife and other ecosystem services as well, with NTFRs still a low priority. Native American tribes with reservations are faced with pressure from the growing commercial NTFR industry and its opportunities and liabilities. The wide variety of access systems utilized by landowners in the region span a wide range of systems and access requirements. Some permits are short term, lasting only two weeks, while other landowners require a three-year lease. Some permits only cost US$20, while a lease might cost US$10,000. Some permits are sold to an unlimited number of people for a given area; others are specified for only one family. Some land managers only require a Mexican driver's license as identification; others require a business license, contractor's license and proof of insurance. Some tribes will allow non-tribal members to harvest on tribal lands for a fee; others will not.

Salal harvesters can feasibly make a good income if they have their own transportation into the forest and can negotiate with landowners for permits and for good prices from buyers. However, many end up paying for transportation from a driver, paying an inflated price for a permit and giving a percentage of the day's product to a 'patron' who holds the permit and the connection to a buyer. Given these conditions, 'poaching' or un-permitted harvesting occurs regularly on both private and public lands, often overwhelming any planned management practices on the part of land managers and harvesters with permits. It is within this management context that harvesters make decisions about where, how much and how to harvest floral greens on the Olympic Peninsula.

Comparative case studies: Two communities of NTFR users on the Olympic Peninsula

The complex social, ecological and economic dynamics of floral greens harvesting on the Olympic Peninsula provide the context for a comparison of two projects initiated in 2001 and 2002 using a participatory research approach to study the sustainability of salal harvesting: Case 1 was carried out with floral greens harvesters in Mason County, Washington, and case 2 with the Makah Indian Tribe's

Community-based Forestry Initiative in Neah Bay, Washington. These communities are located a four-hour drive away from each other on the Olympic Peninsula, separated by the Olympic Mountains. Prior to the initiation of the research projects, the Makah Indian tribal members and the floral greens harvesters from Mason County rarely interacted. However, as both groups have pursued economic development activities based on NTFRs, some members of the communities have begun working together on training and planning projects.

Despite similarities in both the motivation and ecological knowledge to manage salal sustainably on the part of some members of the community, there are significant differences in the communities. The floral greens harvesters in Mason County and Neah Bay both have a strong interest in maintaining the sustainability of NTFR harvest, though often for different reasons. Both groups have the potential to provide valuable information and strategies for not only managing NTFR resources, but for researching and monitoring these resources as well.

Background on floral greens harvesters in Mason County and the Makah Indian Tribe

Floral greens harvesters in Mason County are predominantly male, primarily migrant workers from Mexico and Latin America and immigrants from Southeast Asia. Most floral greens harvesters have been in Mason County harvesting for fewer than ten years and do not own their own land. Because harvest research specifically relies on ecological knowledge and harvest practices for a particular species, the harvester community is defined as anyone who picks and sells salal commercially for any part of the year. Long-time Anglo resident harvesters whose parents and grandparents taught them how to pick a variety of NTFRs make up a very small part of the harvester community; but most of the current harvester community does not have this historical connection to the land and is generally not considered part of the local community in the towns and forests of the Pacific Northwest. However, the particular livelihood practices of floral greens harvesters provide them with the opportunity to gain extensive local ecological knowledge of, and familiarity with, the forests in which they work.

Most harvesters who pick salal do so for the majority of their income, but also pick a variety of other NTFRs when their prices are high, when wholesalers need workers and pay well during the Christmas season and during salal's off season (May and June). Many harvesters from Mexico and Guatemala return to their home country for one or several months to visit family or work on family farms or ranches. Many harvesters have only a few family members, primarily brothers and cousins, on the Olympic Peninsula. The majority of floral greens harvesters on the Olympic Peninsula are thought to be undocumented workers (Hansis, 2002; McLain and Lynch, 2002). Picking floral greens seems to be a fairly low-profile, entry-level job in which a person with few English-speaking skills can obtain a permit with no documentation (or not obtain a permit at all), find transportation into the woods and sell his or her product each day for cash, all under less threat of deportation than in many other types of work (though fear of deportation is common and continual). Because many harvesters are undocumented and speak little English, their access to either government assistance or non-governmental organization (NGO)-sponsored economic support programs is very limited. Furthermore, because much of the industry is unregulated and occurs 'under the table', many harvesters are vulnerable to exploitation by wholesalers. Similarly, harvesters have little to no access to, or influence on, formal forest management practices despite their direct impact on the resource and their experience with the resource (see Table 9.1).

In 2001, the Northwest Research and Harvester Association (NRHA), led by co-author Don Collins, formed a partnership with co-author Heidi Ballard, a graduate student at University of California, Berkeley, at the time, to design and implement harvest impact studies on salal, as well as to help harvesters document their resource knowledge and develop an effective methodology to monitor their harvesting and inventory the resource. An organization founded by and for NTFR harvesters, the NRHA places an emphasis on increasing access to harvestable land for harvesters, as well as training, monitoring and research on both the harvesting and the natural resources themselves (Ballard et al, 2002).

Since its formation in October 2001, the NRHA has had a contract with the Washington Department of Natural Resources (WA DNR) to manage the understory of 16,000ha (40,000 acres) of State Forest in Mason County, in addition to several agreements with private industrial timber companies to manage their lands for NTFPs. Hence, as an alternative to the plight of the individual harvester who may not have the necessary skills or power, the NRHA can collectively bargain with landowners and wholesalers for better land access and better prices for their product. Specifically, the stated purpose of the agreement is to conduct a cooperative research project within the two state forests in order to inventory understory plants and to help integrate NTFR management regimes with timber and stand management regimes.

The Makah Indian Tribe, which consists of five associated villages that existed before the Treaty of 1886 created the Makah Reservation, has occupied the area of the remote northwest Olympic Peninsula on which their reservation now resides for thousands of years (archeological evidence of their gray whale hunting dates back 2000 years; Renker, 1997). They are a Nootka band more closely related linguistically and culturally to the First Nations tribes of Vancouver Island, British Columbia, than to other tribes on the Olympic Peninsula. The current population of the reservation is predominantly Makah tribal members and a small percentage of non-tribal members. Approximately 5 per cent of the approximately 12,000ha (30,000 acre) reservation is private allotments and the remainder is tribal land designated by the Makah Tribal Council as a protected area for wilderness, a protected area for cultural purposes, or an area used for timber production (Makah

Table 9.1 *Comparison of community characteristics of the salal harvesters of Mason County and the Makah Indian Tribe*

Community characteristics	*Mason County salal harvesters*	*Makah Indian Tribe*
1 Duration of connection to the land	Most have arrived in the area within the last 20 years.	Families have occupied this land for several thousand years
2 Economic dependence upon non-timber forest resources (NTFRs)	Most derive the majority of their annual income from the harvest and sale of NTFRs, primarily floral greens.	Few harvest floral greens, but many depend on other NTFRs for crafts and other products for supplementary income.
3 Cultural and family responsibilities in the immediate area	Most have few family members in the area.	Many devote time and energy to cultural practices such as ceremonies, dinners and care of family members.
4 Access to/influence on formal forest management practices	There is little or no access (especially by the many undocumented, non-English speaking workers).	Tribal members may vote on forest management plan and have direct access to forest managers.
5 Access to government- and NGO-sponsored economic support programs	There is little or no access because of undocumented worker status and a lack of awareness of services for agricultural laborers.	A variety of services and support are available for housing, monthly income, food, energy and health assistance.
6 Overall mobility and continuity within the community, both spatially and temporally	Many move from job to job seasonally and/or during a harvest season, and may or may not return the following year.	Most community members remain on the reservation year round, many for their lifetime.
7 Specific goals for the NTFR research	Goals are to support the agreement with the land managers to research and monitor NTFR harvest; to learn about the impacts of harvest practices that they already employ and to learn research methods and skills in order to conduct more research projects in the future.	Goals are to conduct a pilot project to learn about commercial NTFR harvest impacts and economic development potential; and to train community members in forestry and natural resource data-collection skills.

Indian Tribe Forestry Program, 1999). This community's identity, in contrast to floral greens harvesters in Mason County, is much more defined by a historical connection to the land, shared culture and ethnicity, and a sense of place (see Table 9.1).

With respect to economic dependence upon floral greens harvesting, very few people, if any, depend solely upon salal harvesting, though many tribal members supplement their income with other NTFRs, such as red cedar and firewood cutting, red cedar bark basket-weaving or wood carving. A large portion of the community also directly or indirectly benefits from the harvesting of numerous plants used for medicines, alder for smoking salmon and berries for food. The two most prevalent livelihood practices are fishing for salmon, halibut, black cod, whiting and other fisheries, or working for the tribal government in a variety of government service positions. In contrast to the mobility of the floral greens harvesters in Mason County, the Makah community members are much more stationary, working year round on or near the reservation or on boats in the adjacent fisheries. Though fishermen's incomes can be very high seasonally, many of these people are unemployed for the rest of the year, when supplementary income sources from the forest, as mentioned above, and bartering become more important. Extended and tight family and cultural support networks within the community are an important factor in relation to these livelihood strategies and also require extensive time and energy of tribal members. Community gatherings, traditional dinners, traditional dances, fundraisers, family meetings and other responsibilities play a large role in the lives of many community members, which contrasts with some floral greens harvesters, and can play a significant role in a participatory research project on the reservation.

In terms of access to government- and NGO-sponsored economic support programs, members of the Makah Indian Tribe may be eligible to draw on a variety of federal, state and tribal programs that provide monthly general assistance, food commodities, energy, Indian Health Services and housing. While many tribal members are employed and do not benefit from all of these programs, some members can and do utilize them for all or part of each year. In addition, the tribe as an organization is capable of securing NGO funding for economic development projects, conservation and environmental project funding, all of which can be used to fund research and management of NTFRs and often encourage participatory and/or traditional resource management practices, such as the project described in this chapter. Finally, in terms of access to, and influence on, formal forest management practices, the long-term ownership and treaty rights of the Makah Indian Tribe mean that tribal members have access to much of the land and resources of the reservation, though some areas are closed for wildlife or cultural protection. The structure of the tribal government means that individuals may vote on and also have direct input into forest management and use policies. All of these factors, many in contrast to the community of floral greens harvesters of Mason County, seem to provide a very rich and receptive context in which participatory research on non-timber forest resources could take place (see Table 9.1).

With the overall goals of ensuring that the forests remain healthy and productive for future generations, the Makah Indian Tribe created the Makah Community-based Forestry Initiative (MCBFI) – part of the Ford Foundation's Five-year Community Forestry Demonstration Program – overseen by Forestry Program Manager and co-author Joyce Trettevick. The goal of the MCBFI was to partner with the timber resources management section to manage NTFRs on the reservation and to 'help develop "holistic" management policies and regulations that promote the restoration, protection and economic diversification of the forests for optimum environmental, cultural and economic benefit' (MCBFI, 2002). As part of these goals, the MCBFI project facilitated the sharing and enhancing of community members' knowledge and skills in sustainable NTFR management practices. For products that are already harvested by tribal members, intergenerational teaching and learning are enhanced through sustainable harvest workshops that draw on expertise within the community. However, floral greens such as salal had not been widely harvested by tribal members for commercial sale.

In 2002, Ballard was approached by the MCBFI to:

- develop a pilot study on the commercial harvesting of salal on the reservation; and
- train community members with some of the skills to be forestry and natural resources field assistants.

Because very little is known about sustainable levels of harvest of floral greens species, the MCBFI Steering Committee decided that small-scale harvest experiments are needed to make sure that the forest understory which the Makah have depended upon for thousands of years remains productive and diverse far into the future. Significantly, the salal harvest pilot study for the Makah was not initiated by people familiar with salal harvesting, but by the MCBFI Steering Committee, whose goal was to explore the possibility of promoting floral greens harvest on the reservation as a livelihood option for tribal members. The steering committee consisted of three professionals employed in full-time jobs with the tribal government and three 'community members', one a self-employed carver, one contract employee of the tribal government and one self-described 'homemaker' dependent upon seasonal jobs. Also significant is the fact that the research project was initiated by the steering committee, not by Ballard, nor by tribal members who would hopefully benefit from this livelihood option.

THE PROCESS: APPLYING A PARTICIPATORY RESEARCH APPROACH FOR SALAL HARVEST

Case 1: Mason County salal harvesters

Problem identification

The Olympic Peninsula Salal Sustainability and Management Research Project (hereafter called the Salal Project) was initiated in 2001 by both the Northwest Research and Harvester Association (NRHA) and Ballard, and continued through

2004. The main objective of this project was to collaborate with harvesters and land managers to design and conduct experiments testing salal harvesting techniques.

Consequently, Ballard conducted participant observation and semi-structured interviews in 2001 and 2002 on the Olympic Peninsula with salal harvesters to document their understanding of the ecological dynamics of the resources as a basis for participation in research and management. The collaboration began with Collins suggesting that the focus of the project should be on salal harvest sustainability. Ballard initially provided the structure and logistics of the research process; but harvesters from the NRHA were invested and involved in some way in every stage of the project since.

Determining the research question

The project began with interviews of salal harvesters who described very tight connections between the resource tenure regimes and harvesting practices. In many cases, when asked to describe their harvesting practices, harvesters replied: 'It depends on whose land I'm picking on. If I know I'm never coming back to an area, I'll pick it different from when I know I'll come back next year.' This made the case for a participatory approach to the ecological research even stronger because harvesters explained that they choose to harvest in different ways depending upon the type of land access they have. Before the collaboration with the NRHA, Ballard assumed that the incentives for, and effects of, harvesting different commercial grades of salal were the most important questions to answer regarding salal sustainability. However, the entire trajectory of the question changed after collaborating with the NRHA. Harvesters helped to design the research question such that the two harvest practices most often described, which seemed to correspond to either open-access/short-term access situations ('heavy harvest practices') or long-term access/direct relationship to landowner situations ('light harvest practices'), were tested in the harvest experiments. Therefore, Ballard framed the research question to focus on a controlled salal harvest experiment; but the harvesters determined the specific questions regarding harvest intensity. While Collins and some of the more experienced harvesters understood the longer-term (three-year) trajectory of the project, other harvesters did not necessarily feel ownership over the question or project, saying:

> *... many friends, many fellow pickers were surprised: they thought we were crazy because they had never been near or interacted with this kind of study, and I tell you, it is the first experiment that has been done and I said that ... I would try it out, test it out. It is a requirement from the university for [Ballard]. It is not for my benefit; but maybe with time it will be and will be even more grateful; but we will work with her. Maybe we will see it in a few days or in 15 years.*

Experimental design

Salal harvesters participated in all aspects of designing and conducting the salal harvest experiment. After many discussions with other plant ecologists and harvesters, the experimental design that had relied solely on a review of the literature and discussions with ecologists was revamped. Research site locations chosen by harvesters reflected the variety of environmental conditions in the area, such as differing elevations and forest stand types. Most importantly, harvesters and Ballard collaborated to design the particular response variables in the plant that would detect impacts due to harvest; these were variables not found in the literature that could more specifically measure impacts of harvest. Harvesters and Ballard also carefully defined the harvest treatments to be tested based on the actual 'light' and 'heavy' harvest practices, combining real-world practices with guidelines of biomass removal experiments. Methods from the literature had to be modified to encompass the parts of the plant actually harvested, and harvesters had to compromise efficiency of harvesting for thoroughness and uniformity of biomass removal. Because the ecology and management of NTFRs are relatively new pursuits for ecologists in the US, harvesters' expertise and experience with the methods and effects of harvest on salal were essential in bringing these particular components of NTFR research to the project. Complementary to the harvesters' contributions were the statistically sound experimental design and plant ecology field methods that Ballard provided.

Data collection

Over the course of three summers (2001 to 2003), Ballard trained approximately six harvesters as field assistants for the data collection. Harvesters were paid US$10 to US$12 per hour, which was considered to be more than they made picking salal, to spend anywhere from two to eight hours a day for a varying number of days collecting and recording data with the researcher. Contrasting with a typical ecology field season in which a set number of field assistants would be trained and then continue to work the entire season (and sometimes several seasons), the Salal Project had to remain fluid such that several new harvesters needed to be trained as others became unavailable. In addition, four US Forest Service technicians from the Hood Canal Ranger District of the Olympic National Forest were also trained and collected data, often in teams with salal harvester field assistants. The data sheet was translated and data was collected in Spanish so that harvesters who spoke very little English could participate. Finally, harvesters determined and applied the harvest treatments: a light intensity (33 per cent removal) and heavy intensity (100 per cent removal) of the commercial-quality salal, weighing and taking samples of the product harvested.

Interpretation of data

One of the more difficult stages of research to truly incorporate community participation is the analysis and interpretation of data (Minkler and Wallerstein, 2003). In the case of the Salal Project, harvesters did not have the expertise to conduct the statistical analysis, so Ballard compiled the results. However, in September 2003, 20 harvesters in the NRHA gathered to interpret the harvest yield results for each year for each experimental site using large bar graph representations of the results. With instructions and discussions on how to read bar graphs and several harvesters serving as Spanish translators and facilitators of small group discussions, harvesters discussed why some results differed from their hypotheses, why sites responded differently to the same harvest treatments and how the results could be used for management recommendations. Crucially, several harvesters pointed out how environmental conditions might have affected the results, one man saying: 'You know how it was a really dry year last year, and so some areas had more bugs (eating the leaves) than others. Maybe you got different (yields) for different areas because the bugs ruined (the quality) of some areas.' These were directly applied to the management recommendations produced by the project.

Drawing conclusions and dissemination of results

Ballard wrote the dissertation in 2004, and the results of the salal harvest study were given to Collins, as well as the United States Forest Service (USFS) district vegetation manager, the Washington State DNR land managers and foresters, and the private timber company NTFP coordinator where research sites were located. These included recommendations for permitting and management changes based on the results of the salal harvest experiment and on input from harvesters in the NRHA. Land managers said that they hoped to use the results of the study for forest management decision-making and many expressed surprise at how knowledgeable harvesters were. In retrospect, the results were not well disseminated to the harvesters within the NRHA. Most of the members, whose first language is Spanish, needed more of the project results translated. While plans were made to produce a short report outlining the results of the research to be translated into Spanish, time and resources restricted the distribution of information to a brief presentation by Collins at a meeting of the harvesters.

Several harvesters of the NRHA continued to conduct this kind of research on a different NTFR species with another researcher, which was an important goal of the project from the start. This type of community capacity-building is also a goal of participatory research as a tool for social change, and can empower harvesters to participate more effectively in the forest management dialogue (Fawcett et al 2003; Israel et al, 2003).

After the first year of the Salal Project, Ballard was approached by a neighboring community who was also interested in conducting research on the impacts of harvesting salal: the Makah Indian Tribe. In contrast to the NRHA harvesters, they were seemingly unhampered by the obstacles of access to land and to forest management decisions because they control and manage their own forest lands.

Case 2: The Makah Community-based Forestry Initiative Salal Pilot Study

Problem definition

The MCBFI approached Ballard precisely because of the project that had already begun in Mason County during the previous year. However, the MCBFI identified the problem in a different way compared to the floral greens harvesters, which ultimately affected the way in which the research progressed. Whereas the salal harvesters in Mason County knew that large-scale commercial salal harvesting was already taking place on their leased land and wanted to test different harvest intensities, the MCBFI wanted to determine whether the tribe should allow harvesting at all by testing its commercial and ecological sustainability. Importantly, an additional goal of the MCBFI was a commitment to providing training and job skills to as many tribal members as possible through the process of the experimental research.

Defining the research question

In contrast to the harvesters' Salal Project, the research question was not developed by the MCBFI Steering Committee, but was borrowed from the Salal Project developed by the floral greens harvesters and Ballard the year before. They wanted to apply the same question on their own land with their own community members. Therefore, the question remained one focused on determining the effects of different harvest intensities on the salal. However, as stated above, the question was being asked for different reasons than with the floral greens harvesters, and Makah community members were significantly less involved in its development.

Experimental design

As was true for defining the research question, the initial experimental design had been determined the previous year with the salal harvesters in Mason County. Although the MCBFI Steering Committee had assumed that the same experimental design could be used on the Makah Indian Reservation, the cultural and environmental context demanded that the design be modified. This was primarily because the tribal members participating in the project had different kinds of ecological knowledge than the salal harvesters had – that is, most did not have experience with the harvesting and management of this particular species. The cases in Mason County and the Makah Reservation were similar in that the harvesters participated in the placement of the experimental sites within the landscape. They differed, however, in that the Makah tribal members involved had varying levels of experience in the forest and so used different criteria to chose experimental sites. For example, because one member was a woodcutter, he chose a place that he knew would not be damaged by woodcutters, with less of an emphasis on the quality of the salal there. Through this process, in the summer of 2002, three experimental sites were installed in stands of varying age classes, elevations and accessibility. This process included the informal on-the-ground training of five community members for the experimental design and layout, who were hired by the MCBFI project; the process occurred over the course of two months.

Data collection

Ballard trained several more tribal members in data collection methods and then conducted pre-treatment data collection on regrowth variables on sub-plots within the three experimental sites on the reservation. This included a two-day workshop on plant inventory and monitoring field methods offered to anyone on the reservation who could then opt to apply for a field assistant position with the MCBFI project. For both pre- and post-treatment data collection, the same plant growth variables as for the NRHA Salal Project were recorded within the three different experimental sites. In November 2002, the first harvest treatments were applied to all three sites, which included the training of seven more community members to harvest salal at different levels of intensity Ballard and by one of the few community members who had experience harvesting floral greens. This was in sharp contrast to the Mason County case, in which participating harvesters defined the treatments and had to train Ballard in harvesting methods. In September and October 2003, two tribal members were trained to collect the post-treatment data without the university researcher present, beginning the transition of the project to the complete control of the MCBFI participants.

Interpretation of data, drawing conclusions and dissemination of results

The goal of the project was to train tribal members to continue the research without Ballard. However, the MCBFI Steering Committee decided that continuing the project was not an efficient use of funds. This is likely the result of the many lessons learned during the pilot study, not as much about salal harvest impacts but about the questionable appropriateness and feasibility of a larger-scale salal harvest program on the Makah Reservation. The plan for the pilot study was for the results to be interpreted in collaboration with the MCBFI project team and then integrated within the Makah Forest Management Plan. Yield and abundance results compiled from the first year of data collection were provided to the Makah Forestry Program. The Salal Harvest Pilot Study will be used to develop the management plan for floral greens on the reservation if large-scale commercial harvest is allowed, which will include specifications of harvest levels. Several community members who participated in the pilot study subsequently provided workshops on sustainable harvesting practices for floral greens for community members and have since started their own salal-buying 'shed' off reservation.

DISCUSSION: EFFECTS OF PARTICIPATORY RESEARCH ON COMMUNITIES, RESEARCH AND THE MANAGEMENT OF NATURAL RESOURCES

Understanding that participation is 'a goal and a process' (Greenwood et al, 1993), both the Salal Project with Mason County harvesters and the pilot study with the Makah Tribe began with the goal of involving people local to the resource as much as possible in the ecological research of the impacts of salal harvesting. However, neither project satisfied all of the principles of participatory research outlined by Israel et al (1998). A comparison of the two projects using these principles reveals the extent to which meaningful participation at every stage of the research was actually achieved and in what areas it was lacking (see Table 9.2). While this comparison reflects the constraints inherent in the contexts of each project, the most important lessons come from an analysis of the outcomes of the participatory research approach, based on the assumptions of the impacts that local participation will have on the community, on the research and on the policy and management related to the sustainability of the resource itself.

Effects on the community

Salal harvesters of Mason County

Migrant and immigrant natural resource workers in the Pacific Northwest have been described as the 'invisible mobile workforce', constituting a large portion of local forest-dependent communities, while still not considered part of the 'local' community. Since the initiation of the Salal Project, harvesters have worked alongside US Forest Service field technicians in the woods, discussed forest policy and management with USFS and Washington State DNR land managers, presented and demonstrated harvest techniques to community forestry scholars and practitioners from throughout the US, discussed ways to informally collaborate with other non-profit organizations working to improve the livelihoods of mobile workers, and collaborated with DNR researchers on a mushroom harvest monitoring study and white pine blister rust mitigation research. These interactions with agencies and other members of the community occurred despite the fact that harvesters did not fully participate and have control over every stage of the research (see Table 9.2, no 3). Hence, it may be that participation in every stage of the research is not as important as meaningful interactions between the community participants – in this case, harvesters and the more powerful local elite who may provide better

Table 9.2 *Principles of participatory research in the case of the Makah Salal Pilot Project*

Principles of participatory research	*The process of the Salal Project of the Northwest Research and Harvester Association (NRHA)*	*The process of the Salal Pilot Project of the Makah Community-based Forestry Initiative (MCBFI)*
1 Community-driven topic	• Research topic directly addresses livelihood practices of the harvester community.	• Research was an exploratory pilot project of interest to a few community members.
2 Builds on resources in the community	• Harvester local ecological knowledge was an essential foundation for the ecological experiment.	• The project focused on a commercial product, drawing on the experience of some community members with commercial non-timber forest products (NTFPs), but only cursorily drew on traditional ecological knowledge of the community.
3 Community members participate in all phases of the research	• There is equitable participation in defining the research question, experimental design and data collection. • The harvester community is involved; but problem identification, data analysis, interpretation of results and dissemination of information are controlled by the university researcher.	• Makah community members were involved in problem identification, definition of the research question, experimental design and data collection; but, ultimately, everything was controlled by the university researcher. • There was no involvement of community members in data analysis and the interpretation of results.
4 Integrates knowledge and action for mutual benefit of all partners	• Research results will be used to inform management affecting harvesters' livelihoods. • Harvesters benefit from relationship with university researcher by receiving technical assistance for other NRHA activities; university researcher benefits from relationship with harvesters via improved research and publications.	• Research will be used to inform formal forest management practices affecting NTFRs by the Makah Indian Tribe. • The community benefits from training a few Makah tribal members for natural resource work;the university researcher benefits from improved research.
5 Promotes co-learning and empowerment	• University researcher learned about harvest from the harvester community in order to conduct the research; harvesters learned about scientific method and data collection to conduct the research. • Forest Service managers learned of harvesters' extensive management knowledge, began to consult harvesters on NTFR management issues.	• Participating Makah tribal members learned about harvest from the university researcher; however, because trust was often lacking, co-learning was minimal. • The project brought the concerns of some community members about NTFR management to the attention of tribal foresters.
6 Cyclical and iterative	• Distance of university researcher from harvester community and mobility of harvesters made trust and partnership building slow and interrupted, but feasible.	• A variety of obstacles to trust and partnership-building resulted in a limited scope and application of the pilot project.
7 Disseminates findings to all partners	• Final dissertation was given to local forest managers and NRHA leaders; the process and results were published in books and peer-review journals. No real dissemination to harvesters occurred.	• University researcher presented the preliminary findings to the Makah Forestry Program and the MCBFI, but not to the broader community.

livelihood conditions for the community. The interactions with agency staff and landowners not only increased awareness of harvesters as contributing members of the community, but were informal exhibitions of harvesters' knowledge of forest ecology and management. This increased harvesters' voice in discussions of forest management has created more formal avenues for participation than they had prior to the participatory research project. For example, several harvesters in the NRHA have attested to improved relationships with landowners, managers, buyers and researchers. One WA DNR manager stated: 'Things are so much better now that harvesters wave at me with their whole hand instead of running in the other direction. It really helps us out that they know the forest around here so well.' Additionally, as a result of working on the Salal Project, one harvester began the process of pursuing a field technician position with the Forest Service. Connections like these show promise that the process of participatory research has the potential to improve relationships with forest management agencies.

Evidence of the positive impact of participation by harvesters in the Salal Project can be found in the collaboration between the NRHA and a Washington State Department of Natural Resources forest pathologist studying white pine blister rust. By truly drawing on resources and knowledge already in the community (see Table 9.2, no 2), the Salal Project became a template for how harvesters could engage with researchers on the land that they manage. In 2003, the NRHA and Washington State University Extension initiated a project with the pathologist to test the effects of bough harvesting as a potential for mitigating the spread of rust in white pine trees. The results will be used both by the NRHA in their management practices and as a tool for negotiation with landowners and by the WA DNR in their land management policies and practices state wide. One harvester commented:

> *To tell the truth, I had never been so close to a biologist, working with one, until two years ago [when he began working with Ballard], and I got a lot of experience to be able to move forward and be able to support someone else if there was to come another student about to finish their career/studies, be able to support them and get a bit more practice and a little more science.*

This continuation of harvesters' and researchers' joint production of scientific knowledge beyond the confines of the Salal Project represents a small positive move towards more democratic scientific research.

The Makah Indian Tribe

Participation by the whole community in all phases of the research process was only partly achieved during the Salal Harvest Pilot Study (see Table 9.2, no 3). In contrast to the harvesters' Salal Project, the Makah tribal members who helped to locate sites and collected data were not involved in defining the problem and research question from the outset (see Table 9.2, no 1). Furthermore, there were few opportunities during the project for the community members to interact with the local elite – in this case, decision-makers in the tribal government. Hence, there were fairly few cases in which the most marginalized members of the community were empowered by involvement in the project. However, those few community members who made an effort to work with the Forestry Department staff and learn about the salal harvesting industry have moved to higher decision-making levels in both the government and in business. In addition, there were some positive impacts on the community as a result of the participatory approach to NTFR research. Because this work was not simply conducted by the Makah Forestry Program staff and instead sought community participation through putting up flyers, submitting articles to the tribal newsletter, conducting training workshops and employing a variety of tribal members, the project increased awareness and dialogue about NTFRs and their users, past, present and future. The project in a small way did promote co-learning and empowerment (see Table 9.2, no 5) by promoting productive dialogue about the appropriate uses of the forest for current and future generations between tribal government officials and community members who harvest NTFRs for cultural and subsistence purposes, such as for basketry materials, medicinal plants, edible berries, mushrooms and herbs, and fishing and construction materials. Tribal members who wanted improved infrastructure for commercial harvest of NTFRs debated with those concerned about reports of over-harvesting elsewhere on the Olympic Peninsula. This dialogue was crucial to advancing the tribe's exploration of NTFRs as a possible source of income for the poorest members of the community.

Throughout the process of recruiting, training, designing and implementing the project, the MCBFI learned about who in the community has an interest in harvesting raw product such as salal and other floral greens, as well as the level of interest in natural resource-related jobs as a whole. Contrary to the preliminary assumptions of many of the steering committee members, because all the examples of successful sustainable salal harvesting were from off the reservation and from very different ethnic and socio-economic groups, few community members saw salal harvesting as a viable full-time job alternative. There was, however, interest from several community members in gaining harvesting skills in order to supplement their income on a part-time basis. One participant said after being trained to harvest and after applying the treatments: 'I didn't realize it was so easy. Now that I know how to harvest, I can go out any time I want … but I think I would only go out once in a while when I need extra money.' The process of attempting participatory ecological research, therefore, provided a vehicle for communicating the needs and interests of often unheard members of the community to the tribal natural resource managers. Hence, while the project did not satisfy many of the principles of participatory research or empower the disempowered community members (see Table 9.2), the process of involving the community allowed for frank discussion between community members about forest management priorities that might not have occurred otherwise.

Effects on the research

Salal harvesters of Mason County

Harvesters contributed substantially to the experimental design of the Salal Project in Mason County. It would have been of limited practical value for an ecologist to experimentally harvest salal in the standardized ways of the discipline without consulting harvesters because the results of such an experiment would not reflect real-world harvest situations and would not be applicable to impacts that landowners and harvesters actually face. In fact, at the start of the Salal Project, many ecologists suggested that such harvest experiments should occur in a greenhouse under controlled conditions. While this is certainly the most effective way of reducing variation, controlling environmental and anthropogenic variables, it would completely ignore the real harvest practices used by harvesters across the Olympic Peninsula. Instead, harvesters helped to design the research question such that the two harvest practices most often described were tested *in situ* on public and private lands, where harvesting actually occurs, thereby maintaining focus on a community-driven topic as a principle of participatory action research (PAR) (see Table 9.2, no 1). This did mean that experimental control was sacrificed for accuracy of harvest conditions. Appropriately, questions of academic rigor and validity regularly weigh heavily in debates on the potential success or failure of a participatory research approach (Bradbury and Reason, 2003). Depending upon the types of questions being asked, a combination of both highly controlled experiments and harvester-designed experiments should ideally be employed to take advantage of the benefits of both types of research for the purpose of answering management questions.

Each step of the research process involving harvesters required negotiation between the accepted level of rigor for ecological research and the principles of participatory research laid out at the beginning of the study (see Table 9.2). For example, a crucial part of the experimental design was the precise definition of harvest treatments that both accurately represented true harvest practices and satisfied the standards of rigor in ecological research. This required some compromise on both sides. In the most controlled situation, Ballard (and/or an undergraduate field assistant) would have individually applied all the harvest treatments to ensure the maximum possible consistency from site to site and year to year; conversely, in the most participatory situation, harvesters might have conducted workshops in which a consensus about the most typical harvest practices was reached and treatments applied by as many participants as possible. However, in order to achieve enough consistency, but still involve harvester participation, a core group of harvesters and Ballard collaboratively determined the harvest treatments and applied them over the three harvest seasons of the study. Rather than representing a loss of validity, this compromise represents a successful navigation of two sets of standards, resulting in a more valid experimental design and a step closer to enabling harvesters to conduct their own research.

The Makah Indian Tribe

Just as in the discussion of impacts on the community above, many of the intended goals of a participatory approach to the Makah Salal Harvest Pilot Study were not realized. In this case, the impacts on the research were in some instances positive, some negative, but generally neutral. The research question and most of the methods had already been determined by the Salal Project in Mason County and Makah participants had little to no experience harvesting salal, so there was little opportunity for contributions to the research in this part of the process (see Table 9.2, no 3). However, the participating tribal members did improve the research by locating the experimental sites in high-quality commercial salal areas not likely to be compromised by woodcutters looking for cedar. Their participation also improved the research by connecting the project to the community, thereby introducing to the project the few remaining tribal members who did have experience harvesting salal 20 years ago and incorporating the local knowledge of commercial harvest that remained in the community (see Table 9.2, no 4). These women had full-time jobs and could not participate in the day-to-day process of the research; but they provided valuable insight into where and how harvesting previously took place on the reservation. They also suggested how to incorporate more participation from younger tribal members to apply the harvest treatments. These contributions were invaluable for executing the research in the context of the reservation, though not as important for the experimental design and methods.

Effects on the policy and/or management of the resource

Salal harvesters of Mason County

Using a participatory approach to the study of salal harvest impacts resulted in moderately positive outcomes on both policies of NTFR harvest permitting and access, and on management of NTFRs within public land management agencies in Mason County. Recommendations for permitting and management changes based on the results of the salal harvest experiment and on input from harvesters in the NRHA were distributed to local private and public land managers, a key principle of participatory research (see Table 9.2, no 7). However, results of the project were not translated into Spanish as was originally planned; so while agency personnel and landowners learned the results of the research, dissemination to harvesters was much more sporadic (see Table 9.2, no 7). The dissemination of results to the community is a key principle of participatory research because we assume that the resource management practices will change since resource users have learned from the results of the research. In this case, the majority of salal harvesters did not learn the results of the research, so harvest practices outside of the participants in the project are unlikely.

However, it may be just as important that the land management decision-makers who set harvest policies learned the results of the project. In fact, as public and private land managers became involved with the project, they were surprised by the fact that harvesters were intimately knowledgeable about timber management practices and how they affect understory species. Previously, many managers assumed that harvesters either did not know about sustainable management practices and forest ecology or did not care. After the completion of the Salal Project, the Olympic National Forest began inviting and including harvesters in meetings to discuss harvest and permit policies. Areas were reopened for harvest that had been closed and plans began for a collaborative monitoring project between several salal harvesters and the Olympic National Forest. Because information on salal harvesting and its impacts is totally lacking in formal forest management spheres, both the ecological and social recommendations have the potential to contribute to better forest management practices and to improve harvester livelihoods in the region.

Salal harvest policies resulting from the participatory research process are not the only potential effect of this project on forest management practices. The results of the salal sustainability experiment itself could also have management implications for silvicultural practices on public lands. The research provided information on relationships between overstory characteristics and salal commercial quality, largely due to the knowledge and experience contributed by harvesters to the research process. This information will allow public land managers to consider the overall economic and environmental costs and benefits to timber and non-timber product species of various silvicultural treatments, such as thinning, pruning and clear-cutting.

The Makah Indian Tribe

In the case of the Makah Salal Harvest Pilot Project, the impact of the participatory research attempt upon the policy and management of the resource is significant, despite satisfying only a few of the participatory research principles (see Table 9.2). This may be because the project was initiated by the Makah Forestry Program and this program remained committed to the project throughout, even as we learned that fewer tribal members were actually interested in harvesting salal for income as originally assumed. Hence, the principle of focusing on a 'community-driven topic' for the research (Table 9.2, no 1) becomes muddied; the decision-makers in the community valued the project and were able to act on the results, but interest and investment by the broader community were somewhat lacking. Similar to the harvesters' Salal Project, it may be important that the information was utilized by the higher-level decision-makers despite moderate participation by the broader community. Not only the results of the ecological experiment, but also the information about community priorities for NTFR management that resulted from the research process, are being used by Makah Forestry and MCBFI to improve management of the forest and policies governing access by tribal members to forest products. Makah Forestry's goal is to eventually manage jointly for timber and non-timber products on the reservation, and the Program has used the results of the salal harvest experiment to determine locations and quantity of salal harvest should the tribe decide to manage for commercial salal harvesting.

Similarly, the Salal Pilot Study was used as a template for a research project on the sustainability of bough harvesting for Christmas greenery initiated and implemented solely by MCBFI and Makah Forestry. Specifically, this bough harvesting was integrated within the timber management plan where boughs can be salvaged from units slated for timber harvest. Finally, permitting and access policies are being developed for floral and Christmas greenery harvesting on the reservation that more accurately reflect the needs and interests of the community members, based on the information gained during the participatory process. Previous pre-commercial thinning contracts left the felled small trees on the ground as there was no alternative resource use. This resource is now available for bough harvest and, additionally, provides an avenue to improve forest health by reducing fuels on the forest floor. This integration of timber and non-timber product management will benefit the community members and the tribe economically and socially. The outcomes of the participatory research in each of the two cases are summarized in Table 9.3.

Conclusions: Lessons learned about the benefits and challenges of participatory research on non-timber forest resources

Continuity of participants in all phases of the research process

The questions of *who* participates in the research and of where, when, how and how many people participate were major factors determining the outcomes of each case. In the case of the salal harvesters in Mason County, there was relative continuity between the people who initiated the project, defined its questions, designed the experiment, collected the data and interpreted the results. Though several harvesters participated in some stages, but not others (often because they returned to their home country or obtained full-time employment), a core group of approximately eight harvesters participated in every stage. In contrast, there was a distinct difference in the MCBFI case between the community members who initiated and commissioned the research and the community members who implemented the project. The MCBFI Steering Committee participated in

Table 9.3 *Outcomes of participatory research in each of the two salal harvest studies*

Proposed goals of participatory research	*The Salal Project of the Northwest Research and Harvester Association (NRHA)*	*The Salal Pilot Project of the Makah Community-based Forestry Initiative (MCBFI)*
Effects on the community	• Training for community capacity-building • Increased voice in discussions of local and regional forest management • Establishment of a template for continued civil science research projects with agency researchers	• Training for community capacity-building • Increased dialogue between non-timber forest resource (NTFR) users and formal forest management personnel about appropriate uses of the forest
Effects on the research	• Harvester ecological knowledge provided novel experimental design approaches • Applicability of results improved	• Community members provided positive connections and context for the research, but also slowed the research process
Effects on natural resource management and/or policy	• Relevance of research to local and regional forest management issues may increase applicability • Collaboration with harvester community may improve permitting policies and sustainability	• Participation of Makah tribal members in the research will inform integrated timber and non-timber product management by Makah Forestry

defining the problem and the research question, and will likely be responsible for the management recommendations based on the research; but the people who participated in the day-to-day field work of experimental design and data collection were a different group of community members. The steering committee was made up primarily of 'local elite': full-time employees of the tribal government or other local institutions who were working toward the economic development and conservation of resources on the reservation. The on-the-ground participants were from lower socio-economic levels of the community: often unemployed or seasonally employed in other natural resource work, such as fishing or woodcutting, many were artisans who devoted significant amounts of time to their art and most had uncertainties about the feasibility of either full-time work as a field assistant or as a salal harvester.

The lesson learned from this comparison? There must be consistency between the community members who initiate and define the research project and those who implement it on the ground. There appeared to be much more ownership, responsibility and satisfaction concerning the research on the part of the salal harvesters in Mason County compared to the Makah community members.

Continuity in participation by the Mason County harvesters also contributed to empowering the community members. Whereas the Northwest Research and Harvester Association in Mason County has plans for continuing and expanding the salal research project for several years into the future, the future of the Makah Salal Harvest Pilot Study is uncertain. This is partially to be expected, considering that the Makah project was a pilot study and the lessons learned from it may result in a discontinuance of any salal harvest and the project itself. This action as a result of the research would represent a useful contribution of the participatory research; however, the process itself was problematic.

Another lesson may be that an important component of a successful participatory natural resource research project is the role of a strong leader with natural resource management and human resource management skills. In the case of the salal harvesters of Mason County, harvesters are enthusiastic about participating in the NRHA because of the well-known respectability and knowledge of its president. For over 50 years he has been picking all the commercial NTFR's of the region and has been working with harvesters of every ethnicity. Nearly every land manager and floral greens wholesaler in the area knows him and many harvesters know this. Harvesters' willingness to participate in a community organization like the NRHA transferred to the Salal Project partially because the president helped to coordinate it. The trust built between harvesters and Ballard was largely due to both parties' trust in the NRHA president. Conversely, the MCBFI project had four different project coordinators in as many years; two had no natural resource management experience and three were not Makah tribal members. This made an iterative cycle of trust-building with Makah community members difficult for both Ballard and the project staff, and resulted in less continuity in the research project. The challenge, then, for participatory research to continue in either community will be to establish and maintain a collaborative partnership between university researchers and a well-respected knowledgeable leader in the community who holds the trust of both university and community researchers.

It is also important to note that the Makah Indian Tribe and the floral greens harvesters do not operate separately from each other, but, in fact, are intertwined geographically and economically. The Makah community members who have pursued floral greens harvest now interact regularly with Collins and the NRHA to gain advice and contacts in the industry. Activities have included instructional harvesting workshops by NRHA harvesters for Makah Indian Tribal members. This collaboration has improved relations between the two communities and may enhance the forest management of NTFRs in the future.

Presumed predictors of success of a participatory research approach

Many of the factors that would seem on the surface to provide a receptive context for participatory research on NTFRs instead sometimes served as obstacles to particpatory research, demonstrating that particpatory research should not necessarily be a universal prescription for community-based conservation and development projects. This is true in two ways:

1. community characteristics presumed to lend themselves to particpatory research may not be useful predictors for an effective participatory research approach; each community is a unique case whose internal dynamics and history must be incorporated within any project in order to be successful and similarly,
2. even when two communities are interested in the same research and live in the same ecosystem only miles apart, the particpatory research approach used in one community may not necessarily be the best approach for the other community.

While the undocumented, often non-English speaking, multi-ethnic, mobile, landless floral greens harvesters seemingly scattered across the Olympic Peninsula would seem a difficult context for participatory research, in fact the PR approach improved the ecological research and helped to empower some members of the harvester community. In contrast, while the Makah community has a history of use of the plant being researched, formal avenues of access to the institutions that manage their forests, internal and external economic support for research and development of their NTFRs, and a place-based community with a long history on their land, each of these factors played out in unexpected ways in relation to the participatory research approach and did not achieve many of the goals of participatory research. Based on these cases, therefore, the social, economic, ecological and political context in which participatory research is attempted may be as important as the participation itself. Whether 5, 10 or 50 community members participate in a research project at various stages, taking context into consideration and paying particular attention to social relationships will affect the resulting impacts of the participatory research on the community, the research and the resource itself.

In the case of the floral greens harvesters of Mason County, the Salal Project serves as a case example of how a marginalized group not traditionally considered part of the local community *can* contribute to ecological research. This is true for two reasons: because the research was conducted on a question of inherent interest to harvesters based on their livelihood dependence upon the resource and because they have in-depth knowledge of the resource and its ecology, which contributed directly to the research process and experimental design. These two characteristics, therefore, are the factors that should help to determine who participates in knowledge production regarding natural resources in the US, not whether a person is a long-time legal resident of a particular geographic area, as is often the criterion for determining 'stakeholder' or 'local' status. When scientific research carries an inordinate amount of weight in forest policy and management decisions, the Salal Project demonstrates that both the resource and society will benefit when harvesters have access to this type of knowledge production.

References

Alexander, S., Pilz, D., Weber, N., Brown, E. and Rockwell, V. (2002) 'Mushrooms and Money and Trees', *Environmental Management,* vol 30, no 1, pp129–141

Ashby, J. A. and Sperling, L. (1995) 'Institutionalizing participatory, client-driven research and technology development in agriculture', *Development and Change,* vol 26, pp753–770

Ballard, H., Collins, D., Lopez, A. and Freed, J. (2002) 'Harvesting floral greens in western Washington as value-addition: Labor issues and globalization', *Proceedings of the International Association for the Study of Common Property Biennial Meeting,* June 2002, Victoria Falls, Zimbabwe

Barrow, E. and Murphree, M. (2001) 'Community conservation: From concept to practice', in D. Hulme and M. Murhpree (eds) *African Wildlife and Livelihoods: The Promise and Performance of Community Conservation,* James Currey Ltd, Oxford, UK

Bradbury, H. and Reason, P. (2003) 'Issues and choice points for improving the quality of action research', in M. Minkler and N. Wallerstein (eds) *Community-based Participatory Research for Health,* Jossey-Bass Inc, San Francisco, CA

Chambers, R. (1994) 'The origins and practice of participatory rural appraisal', *World Development,* vol 4, no 7, pp953–969

Fawcett, S. B., Schultz, J., Carson, V., Renault, V. and Francisco, V. (2003) 'Using internet based tools to build capacity for community based participatory research and other efforts to promote community health and development', in M. Minkler and N. Wallerstein (eds) *Community-based Participatory Research for Health,* Jossey-Bass Inc, San Francisco, CA

Getz, W. M., Fortmann, L., Cumming, D., Du Toit, J., Hilty, J., Martin, R., Murphree, M., Owen-Smith, N., Starfield, A. and Westphal, M. (1999) 'Sustaining natural and human capital: Villagers and scientists', *Science,* vol 283, no 19, p283

Greenwood, D. J., Whyte, W. F. and Harkavy, I. (1993) 'Participatory research as a process and as a goal', *Human Relations,* vol 46, no 2, p175–192

Hansis, R. (2002) 'Case study, workers in the woods: Confronting rapid change', in E. T. Jones, R. J. McLain and J. Weigand (eds) *Nontimber Forest Products in the United States,* University Press of Kansas, Lawrence, KS

Israel, B., Schultz, A., Parker, E. and Becker, A. (1998) 'Review of community-based research: Assessing partnership approaches to improve public health', *Annual Review of Public Health,* vol 19, pp173–202

Jones, E. T., McLain, R. J. and Weigand, J. (2002) *Nontimber Forest Products in the United States,* University Press of Kansas, Lawrence, KS

Kerns, B. K., Pilz, D., Ballard, H. and Alexander, S. J. (2003) 'Compatible management of understory forest resources and timber', in R. A. Monserud, R. W. Haynes and A. C. Johnson (eds) *Compatible Forest Management*, Kluwer Academic Publishers, Norwell, MA

Kohm, K. A. and Franklin, J. F. (1987) *Creating a Forestry for the 21st Century: The Science of Ecosystem Management*, Island Press, Washington, DC

Kruger, L. E. and Sturtevant, V. E. (2003) 'Divergent paradigms for community inquiry: An argument for including participatory action research', *Understanding Community–Forest Relations*, General Technical Report PNW-GTR-566, US Department of Agriculture Forest Service, Pacific Northwest Research Station, Portland, OR

Little, P. D. (1994) 'The link between local participation and improved conservation', in D. Western and M. Wright (eds) *Natural Connections: Perspectives in Community-based Conservation*, Island Press, Washington, DC

Makah Indian Tribe Forestry Program (1999) *Makah Indian Tribe Forest Management Plan 1999–2009*, Makah Tribal Council, Neah Bay, WA

MCBFI (Makah Community-based Forestry Initiative) (2002) *Project Summary of the Makah Community-based Forestry Initiative*, 20 April 2002, Makah Forestry Program, Makah Tribal Council, Neah Bay, WA

McLain, R. J. and Jones, E. T. (1997) 'Challenging 'community' definitions in sustainable natural resource management: The case of wild mushroom harvesting in the USA', *Gatekeeper Series no SA68*, International Institute for Environment and Development, London

McLain, R. J. and Lee, R. G. (1996) 'Adaptive management – promises and pitfalls', *Environmental Management*, vol 20, no 4, pp437–448

Minkler, M. and Wallerstein, N. (2003) *Community-based Participatory Research for Health*, Jossey-Bass Inc, San Francisco, CA

Molina, R., Vance, N., Weigand, J. F., Pilz, D. and Amaranthus, M. (1998) 'Special forest products: Integrating social, economic and biological considerations into ecosystem management', in K. Kohm and J. Franklin (eds) *Creating a Forestry for the 21st Century: Ecosystem Management in US Forests*, Island Press, Washington, DC

Murphree, M. W. (1993) 'Communities as resource management institutions', *Gatekeeper Series no 36*, International Institute for Environment and Development, London

Park, P. (1993) 'What is participatory research? A theoretical and methodological perspective', in P. Park, M. Brydon-Miller, B. Hall and T. Jackson (eds) *Voices of Change: Participatory Research in the United States and Canada*, Bergin and Garvey, Westport, CT

Savage, M. (1995) 'Pacific northwest special forest products: An industry in transition', *Journal of Forestry*, vol 93, no 3, pp6—11

Schlosser, W., Blatner, K. and Chapman, R. (1991) 'Economic and marketing implications of special forest products harvest in the coastal Pacific Northwest', *Western Journal of Applied Forestry*, vol 6, no 3, pp67–72

Slocum, R. (1995) 'Participation, empowerment and sustainable development', in R. Slocum, L. Wichhart, D. Rocheleau and B. Thomas-Slayter (eds) *Power, Process and Participation*, Intermediate Technology Publications, London

Stoeker, R. (1999) 'Are academics irrelevant? Roles for scholars in participatory research', *American Behavioral Scientist*, vol 42, no 5, pp840–854

Taylor, R. (2001) 'Participatory natural resource monitoring and management: Implications for conservation', in D. Hulme and M. Murphree (eds) *African Wildlife and Livelihoods: The Promise and Performance of Community Conservation*, James Currey Ltd, Oxford, UK

Von Hagen, B. and Fight, R. D. (1999) *Opportunities for Conservation-based Development of Nontimber Forest Products in the Pacific Northwest*, General Technical Report PNW-GTR-473, US Department of Agriculture, Forest Service, Pacific Northwest Research Station, Portland, OR

Western, D. and Wright, M. (1994) *Natural Connections: Perspectives in Community-based Conservation*, Island Press, Washington, DC

10

Battle at the Bridge: Using Participatory Approaches to Develop Community Researchers in Ecological Management[1]

Jonathan Long, Mae Burnette, Delbin Endfield and Candy Lupe

Prelude

In July 2002, a monstrous wildfire scorched most of the watershed above the town of Cibecue on the White Mountain Apache Reservation in Arizona. Clouds began to gather in the late afternoon, signaling that the summer monsoon rains were soon to arrive. An emergency response team composed of technical experts from various federal agencies and the White Mountain Apache Tribe was preparing for the impending floods. One of the team's first proposals was to clear debris from underneath the two bridges that connected the western half of the town to the larger world beyond. The team was concerned that the bridges would be damaged or overtopped if debris from the fire piled up during floods. Much of the debris was composed of sediments that had washed down in the wake of a smaller wildfire six years earlier. After that earlier fire, we had initiated a stream restoration project in the community in our capacity as employees of the Tribe (when capitalized in this chapter, the Tribe refers to the tribal government). As part of this project, co-author Delbin Endfield had guided local schoolchildren in replanting cattails and other native plants in a popular and well-known site at the upper bridge (see Figure 10.1). The products of their work now lay in the path of a bulldozer. One resident of the community, whose opinion was shared by others, declared: 'The stream is more important to us than the bridge. We do need the bridge; but if nature takes the bridge, that's OK. We don't want you to destroy that place. Our kids worked to make it beautiful again.'

The members of the emergency team had experience working with Native American communities (most were employees of the Bureau of Indian Affairs[2]).

Source: Jonathan Long, 12 September 2001

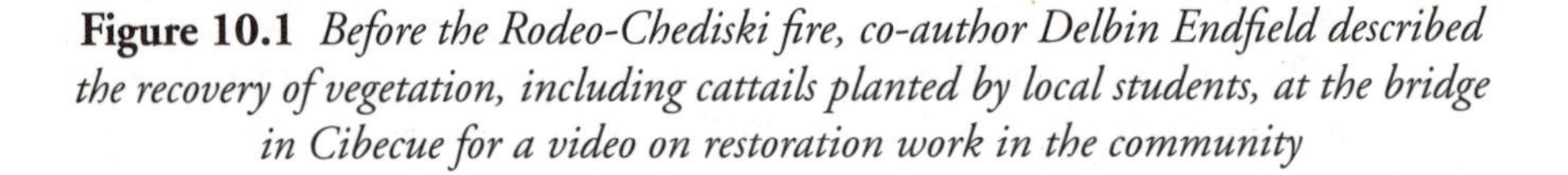

Figure 10.1 *Before the Rodeo-Chediski fire, co-author Delbin Endfield described the recovery of vegetation, including cattails planted by local students, at the bridge in Cibecue for a video on restoration work in the community*

However, the fears of residents were not quelled by the imposition of a command-and-control system staffed by individuals largely unfamiliar to them. Their status as experts in the outside world did not engender a high level of trust among community members. During the fire, one community leader had appealed to community self-reliance by asserting that they should have fought the huge fire themselves using saws and other hand tools. The implication of this statement was that assistance from outsiders was unnecessary and even counterproductive. In viewing the initial emergency stabilization treatments, many residents recalled past ecological destruction directed by outside researchers. In the 1960s, the federal government and the State of Arizona had sponsored an experiment to increase water yield to downstream non-Indian communities by girdling and poisoning culturally significant cottonwood trees along streams in the community (Long, 2000, p228). Seeing bulldozers again preparing to clear vegetation from those streams triggered the community's memory of that traumatic episode. When the local community president called a meeting to discuss the recovery efforts, the plan to clear the creek underneath the bridge became a flash point for debate.

INTRODUCTION

In this chapter, we consider how tensions arise between communities and researchers as they exchange knowledge and resources to manage ecological systems. We draw upon examples from our experiences in developing the Tribe's Watershed Program from 1993 to 2005. We start by briefly summarizing how rural tribal communities such as Cibecue have perceived conventional research, and then we consider how participatory research is intended to yield more reliable information to solve problems. We examine challenges associated with different pathways that have been used to exchange knowledge for problem-solving. The battle at the bridge highlighted the need for the community to have more of its own problem-solvers who would both understand local concerns and be trusted by the residents. However, community members face many of the same constraints that outsiders do, as well as some additional ones. We conclude by outlining participatory strategies to overcome those constraints by broadening local support and capacity to solve ecological problems.

Perceptions of research within communities

Researchers from both universities and government agencies defend research as a means of benefiting society by adding to a common pool of trustworthy knowledge. In many communities, but particularly on tribal reservations, social and ecological problems have proven difficult to ameliorate. This situation has fostered skepticism among community members about the intent and outcomes of research projects. A common perception is that conventional science and research have been a tool touted and wielded largely by outsiders, often against the tribes themselves. One of the most notorious examples of such research involved the use of blood samples from members of the Havasupai in testing human migration theories and the heritability of schizophrenia, when the tribal community had expected help in treating a burgeoning diabetes problem (Dalton, 2004). Tribal leaders can point to such situations as examples of the negative aspects of research; however, they also recognize that research can be used to defend community resources from being taken or degraded by outside interests.

Goals of participatory research

Participatory research and associated methodologies such as action research, community-based research and participatory action research are rooted in the premise that members of a community or organization can and should assume greater responsibility in researching solutions to particular problems, through which they become researchers themselves. A central aim in these approaches is to change power relationships so that historically disadvantaged parts of the community or organization have greater vision and voice in solving their problems (Levin, 1999; Petras and Porpora, 2001, p110). Consequently, participatory research projects strive to make scientific research relevant to the lives of everyday people, rather than serving to increase the power and knowledge of elites.

While cultural traditions provide foundations for management, outside research can provide valuable ideas that can stimulate the evolution of those management systems. By contrasting the roles and perspectives of 'insiders' and 'outsiders', participatory research frameworks help to understand interactions between community members and outside researchers (Elden and Levin, 1991). Insiders have direct knowledge of the organization and are primarily concerned with solving practical problems facing themselves and their organization. Outside researchers may offer expertise, experience, resources and neutrality in conducting experiments, recognizing general patterns and communicating results to others in the research community.

Contrasting worldviews between insiders and outside researchers

The customs and beliefs of insiders and outside researchers often lie in contrast, as is shown in Table 10.1 (with striking parallels to Table 12.2 in Chapter 12 of this volume). In this setting, some issues became characterized as struggles between traditional ways and '*indaa bínatsíkęęs*' ('white people's thinking'). Community members describe how representatives of academic or bureaucratic organizations with formal education often magnify these tensions by using big words to 'show

Table 10.1 *Contrasts between emphases of outsider and insider worldviews*

Outside society and funding entities	*Tribal community and government*
Institutional education transmitted through texts	Traditional knowledge informed by personal life experiences and oral traditions
Written communication in English	Oral communication in native language
Communication with the outside scientific community	Communication with community members
Conceptual knowledge	Practical knowledge
Basic research into general problems	Applications to solve particular problems
Diversity of plants and animals in the ecological community (conservation biology)	Human perspective of the ecological community (land conservation)
Experimental data collection and analysis	Project implementation
Formal reporting about projects	Physical upkeep of projects
Urban lifestyle involving a faster pace and individualism	Rural lifestyle involving a slower pace and collectivism
Maintaining objectivity through distance and appeals to norms of science	Maintaining relationships in the community and upholding traditional values

off' or 'talk down to the people'. Expressing similar frustrations, advocates of participatory research have criticized the notion that universities produce 'expert knowers' or that a scientist's theory about one's world is more valid than one's own (Elden and Levin, 1991; Stringer, 1997). Participatory research seeks to bridge the gap between insiders and outsiders by working together to create a 'local theory' of the situation (Elden and Levin, 1991).

Value of outside knowledge and local research

While community members generally hold traditional knowledge in high regard, they recognize that outside education can help individuals to learn new skills and be more successful in life. For this reason, educational scholarships have constituted one fifth of the annual allocations from the Tribe's permanently endowed Land Restoration Fund. Elders recognized that contemporary ecological research can play an important role in supplementing traditional ways of learning that are in decline. For example, we found that students who have studied plant identification were better prepared to interact with elders who are knowledgeable about traditional plants, even though there are important differences between academic and traditional knowledge. The ability to respectfully adopt new knowledge systems while retaining older ones requires considerable skill in moving between the two worlds. However, such skill can be taught and cultivated through practice.

Modern-day challenges such as rapid population growth, loss of traditional ecological knowledge and climate change have increased the need for local research to search for strategies that fit a particular community's ecological, social and cultural context. Regardless of whether natural resource problems are longstanding or new, solutions require working with a variety of community members who depend upon the land. Particularly in rural watersheds with dispersed populations, command-and-control strategies to address environmental degradation are much more likely to fail than approaches that allow local institutions to adapt through time (Uphoff, 1986).

Background on the Watershed Program

Several factors created an opportunity for us to foster a more participatory community-based approach to watershed management within the 670,000 million hectares (1.66 million acres) of the Tribe's reservation. First, there was strong incentive for the Tribe to build its institutional capacity in natural resources management because a federal agency, the US Fish and Wildlife Service, had threatened to impose regulations regarding conservation of endangered species. Second, another federal agency, the US Environmental Protection Agency, had offered funding for the Tribe to develop a water quality protection program. In addition, the federal government was preparing to compensate the Tribe for past damage to natural resources on the reservation. Co-author Jonathan Long had researched how to respond to these events by conducting interviews with tribal members and non-tribal managers. His analysis concluded that a watershed-based approach could address species conservation and water quality in an integrated manner, while also advancing tribal goals for restoring degraded lands (Long, 1994). The resulting Watershed Program moved quickly beyond the initial objective of assuming authority over water quality protection to conserving and restoring Tribal lands and waters. The Watershed Program quickly grew to comprise eight full-time staff devoted to protecting and restoring the health of the Tribe's water resources.

An early priority of our work was to evaluate how small streams and meadows far from population centers responded to restoration treatments. We learned, however, that a more participatory approach to research was needed when working in tribal communities such as Cibecue. In that town, residents had become highly skeptical of any government proposals for improving their lives (Taylor-Ide and Taylor, 2002, pp178–185). The cottonwood poisoning incident was one of the starker examples of perceived misdeeds at the hands of outsiders who claimed to be helping the local residents. We sought to build local capacity to plan, implement and monitor restoration projects so that they could resist imposition by outsiders.

PATHWAYS FOR EXCHANGING KNOWLEDGE

Apache culture compares knowledge with water, as both are necessary to live well and both can be carried in vessels, since the mind is like a container (Basso, 1996, p134). In Figure 10.2, we represent local knowledge with the traditional pitch-coated wicker basket, or *tus*, while representing non-local knowledge with a metal pail. We use the bridge to represent the transfer of knowledge between the local community and the world of outside researchers. The arrows in the figure represent four different pathways across the bridge.

The first path represents conventional research, in which local knowledge is brought from the community into use or awareness by wider society. The second path signifies the introduction of an outside researcher into the community to address local problems. The third path represents conventional education, in which a community member leaves in order to gain outside knowledge. The fourth path symbolizes efforts to bring outside knowledge to community members within a local educational setting such as community college, extension or similar institutional programs.

Conventional research following the first pathway has helped to record tribal knowledge in lasting, publicly available forms. For instance, many books have recorded Western Apache ecological information in formats that tribal members and resource managers use today. However, community members have criticized

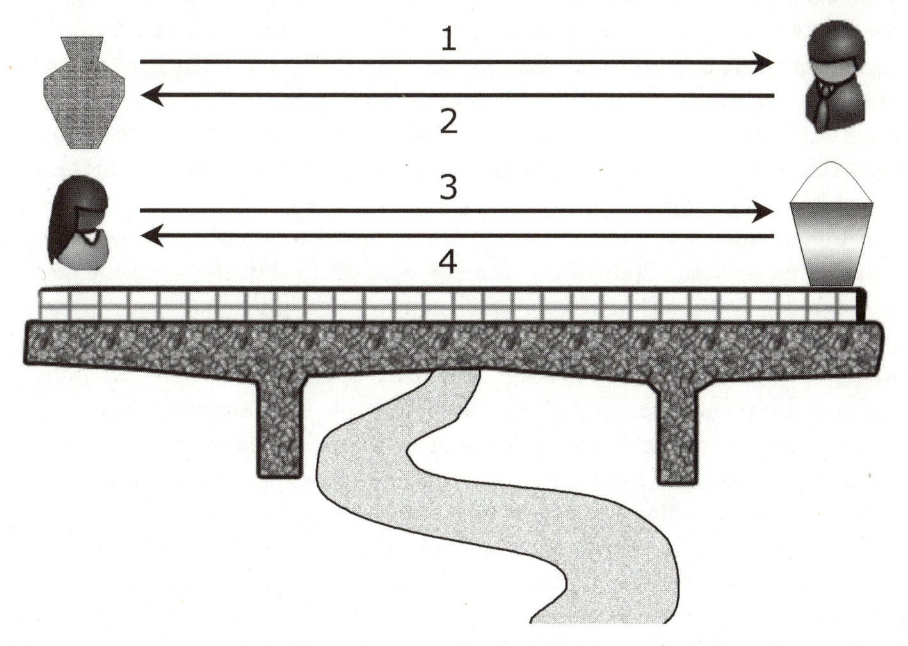

Figure 10.2 *Symbolic representation of four pathways through which insider and outsider knowledge can be transferred between people in the community and those outside the community*

much academic and scientific work as 'continuous probing by outsiders who want answers and knowledge for curiosity's sake, for exploitation, or for research that does not benefit us' (Adley-SantaMaria, 1997, p137). To prevent exportation of knowledge, the Tribe has adopted policies for reviewing proposed research and publications. Such a policy encourages researchers to demonstrate that they are providing some kind of benefits back to the Tribe in exchange for the knowledge that is gained (McDonald et al, 2005). When such a return results in mutual benefits overall, it may be described as a 'parallel process' model (Petras and Porpora, 1993, p111), as resources flow in the two directions.

Other pathways of knowledge exchange can yield sustained benefits to the community. Following the second path, the Tribe has recruited experienced outsiders to perform technical and managerial roles within the government. Persons who have remained for several years have helped to lead many projects and develop local institutions. Some researchers (e.g. Cornell and Kalt, 1995) have attributed part of the Tribe's economic success to its willingness to employ outsiders with specific expertise. Collaborative research projects and involvement in community activities can help those outsiders to better understand the local culture (Ruano, 1991). Outside researchers can become activists as they work with community members to achieve concrete goals (Petras and Porpora, 1993, p119). However, due to the time it takes for an outsider to understand the insider's world, this path is not particularly efficient (Elden and Levin, 1991). For a variety of reasons, outsiders often leave the community after a few years of service. Therefore, bringing in outsider researchers can help to address many short-term technical challenges; but this path is likely to be less effective in addressing more chronic problems, such as ecological degradation.

Some members of the community who followed the third path, by leaving to attend universities, reported that the experience helped them to become more open minded and inquisitive. Upon their return, such individuals can stimulate institutional growth by suggesting new technologies to diagnose or address local problems. Although both outsiders and tribal members who pursue higher education off the reservation can bring in useful tools, community members may be better able to see the trade-offs in adopting a new idea from the perspectives of both an insider and an outsider. Unfortunately, there are many obstacles facing tribal members attending a university. Many of the community members have not been prepared for university-level coursework or for living outside the structure of their home. Furthermore, removing tribal members from their home environment can weaken their ties to the community. Consequently, the few who commit to going to school off-reservation may become even less likely to return. If they do come back, they may be scrutinized to see whether they still subscribe to traditional values. In communities where university education is not a typical path, those who choose it may be viewed as partially rejecting community norms.

The fourth path brings outside education directly to tribal members on the reservation. For example, cooperative extension programs are a well-established way of bringing university research to local communities; but such programs in tribal communities are generally quite young and constrained by funding (Hiller, 2005). The construction of a school or university within the community could be one of the most progressive ways of promoting research to address local problems; but the objectives of the university or the academic world, in general, often take precedence over community needs (Petras and Porpora, 1993, p120). Moreover, obstacles to student enrollment and achievement within conventional academic programs constrain the development of community researchers. The costs of raising young families often make it difficult for individuals to commit the time and money even for locally available community college. Enrollments tend to be low, and dropout rates high, in part because many students are not well prepared in foundational skills such as writing and mathematics. Providing incentives for higher education, such as offering raises when a degree is completed, can have a side effect of seeming to devalue insider knowledge. Furthermore, tribal members who pursue degrees while remaining within the local community may not learn to see the world from a different perspective, as do those who leave the community.

The different pathways for bringing knowledge into a community can address acute problems by importing resources or solutions from the outside; however, they face serious limitations for solving chronic problems. The wildfires in Cibecue were treated as a short-term emergency. However, the wildfires were set by a local firefighter who wanted work to provide money for his family and they grew out of control because of heavy accumulations of forest fuels. Therefore, the damage caused by the wildfires was linked to poverty, lack of employment and past forest management practices. The fires exacerbated these social and ecological problems by prompting the closure of the local sawmill, which had been the lone industry and major employer in the town.

A more complex approach to cultivating knowledge is needed to address chronic problems. We return to the metaphor of the bridge to help explain this problem. Figure 10.2 emphasizes that the bridge serves to transport resources in and out of the community. The bridge is therefore important, but it does not produce anything in itself or improve the quality of the items being conveyed. Instead, we need to consider how the water/knowledge is produced and respected. By focusing on the bridge, the *tus* and the pail in Figure 10.2, we might fail to see that the stream running under the bridge flows from the source of water/knowledge.

Traditional Apache belief holds that spring water is a source of living energy; but when water sits sealed in a container, it stagnates and loses its power. Such stagnation can occur in a sealed *tus* as well as in a metal or plastic container. Living bodies of water dynamically respond to disturbances and can renew themselves. Our systems for building and using knowledge need to be similarly dynamic and self-renewing – they cannot simply harvest, transport and store knowledge. The bridge is an important gathering point where people can share their knowledge; but the stream conveys water from the lands on each side of the bridge. The stream brings new opportunities for discovery, while washing away ideas that no longer thrive. The challenge of community-based research lies in building a bridge that promotes a safe exchange of ideas between the two shores, while respecting the stream that runs through it.

Strategies for resolving tensions on the bridge

Researchers committed to participatory approaches value community knowledge; but they are also likely to believe that outside ideas and perspectives can serve as a catalyst for addressing problems. Efforts to build research capacity need to transcend the model of simply exporting or importing scholars or knowledge. Such one-way transfers (as in 'technology transfer') are not as effective in ensuring that knowledge is valid and useful, as are exchanges based on sharing and dialogue (Phillips, 2006, p233). When realized to its fullest, participatory research promotes collaboration among members with diverse skills and knowledge. It is difficult to design and implement a research effort that integrates community members and researchers. However, successful mixing of inside and outside knowledge could yield answers that are better adapted to the local contexts.

Generating hybridized solutions

Land management activities, such as livestock grazing, agriculture, burning, protection of water resources and erosion control, have longstanding precedents guided by traditional cultural practices (Long et al, 2003). Because local cultural traditions have co-evolved with local ecosystems, they may be more sustainable than management practices imported from other ecosystems. However, many traditional activities require high inputs of labor, and stresses such as population growth and climatic change often make it difficult to sustain traditional ways. On the other hand, novel outside approaches often carry a stigma and are likely to disrupt the social mechanisms that have maintained a system. For example, Merhari (2007) examined efforts by the government in Eritrea to replace traditional irrigation diversions with modern structures. Some of the structures quickly failed due to design weaknesses and inadequate maintenance. Merhari concluded that traditional users should have participated more in the design and upkeep of the irrigation systems.

We addressed the problem of irrigation diversions on the reservation using a more participatory approach. Traditionally, local residents formed teams to construct temporary diversion dams constructed of rocks and brush to irrigate fields along short stretches of rivers (Buskirk, 1986). During the past century, government employees steadily replaced these systems with semi-permanent dams made of rocks, concrete and steel posts. Following severe floods in the early 1990s, failure of these modernized irrigation dams had become a serious problem. In response, we redesigned several of the failed diversions through a collaborative effort with the Apache equipment operators who maintained the diversions, a non-Apache researcher and a non-Apache engineer who oversaw the irrigation program. Our designs sought to emulate boulder or bedrock waterfalls that occur naturally in streams. These structures used native rock materials and induced a 'smoother' flow than the dams that they replaced, so they helped to uphold traditionally important values of naturalness and 'smoothness' (Long et al, 2003). These redesigned structures have performed well since they did not cause any erosion and withstood the abnormally high floods following the wildfire of 2002. Our collaborative approach yielded an outcome that has served the irrigation users, while allowing the rivers to flow more naturally.

Fostering socially acceptable solutions

Another focus of our work was the construction of fences around sensitive springs and wetland areas. In pastoral societies around the world, establishing controls on livestock access is a complex activity because of its implications about rights of access to resources. Many of our early enclosures were broken by elk and horses, or cut open, we suspect, by young people riding all-terrain vehicles and by local ranch hands. These experiences paralleled those in Sahelian Africa, where fencing is 'usually torn down or falls into disrepair because its original purpose is misunderstood or unacceptable to all in the community' (Niamar-Fuller, 1998, p256).

Over several years, we learned how to build fences that would last, both due to better construction to resist wildlife and by gaining cooperation from local community members. We found that sites where fences remained intact tended to not only have high-quality construction, but were also constructed around areas that community members recognized as culturally significant. At one site in Cibecue, local community leaders suggested that we hire a local stockman and his crew of cowboys to build the fence around White Spring, an important water source that had been damaged in a wildfire in 1996. Despite a high level of visitation, that fence remained intact until the wildfire of 2002 burned many of the posts and floods washed out the creek crossings.

At White Spring, the local cowboys built an entry gate from a forked juniper tree, which excluded large animals while allowing people to enter by walking over the crotch of the V-shaped tree. We repeated the design element of the 'V-gate' at other fencing projects. Co-author Mae Burnette, who assumed oversight responsibility for the fencing projects, explained her belief that the V-gate provides a physical passageway for people, as well as spiritual opening for the spring itself. She explained this belief to community members when they expressed concerns about enclosing springs entirely with fencing. Her explanation helped to reassure elders that these projects were being carried out with respect to cultural traditions, rather than as an act of claiming control over the spring.

Promoting translators between traditional and outside knowledge systems

An event in the wake of the wildfire of 2002 exemplified the importance of individuals who can translate between traditional values and outside knowledge systems. Post-fire erosion threatened a culturally important wetland. A federal implementation leader suggested using metal baskets filled with rocks (gabion baskets) to stabilize the channel at the site; but Mae responded that such a treatment would not be a good solution. For one reason, metal may be considered inappropriate for a cultural site as many cultural ceremonies prohibit the use of metal (e.g. participants remove metal jewelry and eyeglasses). For another reason, Mae had previously observed failures of gabion baskets at nearby locations. Fortunately, the staff of the Watershed Program had been engaged in research with an outside scientist to develop a riffle formation technique that uses native

Source: Jonathan Long (June 1999)

Figure 10.3 *Co-author Candy Lupe and an outside researcher, Alvin Medina, work together to install a riffle formation at a culturally important restoration site*

rock and plant materials (Medina and Long, 2004) (see Figure 10.3). Interviews with cultural advisers had demonstrated that the technique was similar to erosion-control practices used traditionally, based upon its design and its reliance on native materials and human labor (Long et al, 2003). When Mae proposed to use rock riffles as an alternative to the gabion baskets, she could offer both traditional ecological knowledge and scientific evidence to support the approach that she thought was culturally and ecologically more appropriate.

Championing land-focused participatory research

While the roots of participatory research extend from the social and management sciences, applying this approach to ecological investigations can integrate ideas and resources from beyond the community with traditional ways of relating to the land. Participatory research in the US has tended to focus on people's relationships to landscapes, rather than on the land itself. For example, Keith Basso's publications about sense of place in Cibecue (see, for example, Basso, 1996) are largely silent about the ecological significance of changes in the land. Staff members of the

Source: Jonathan Long (August 2003)

Figure 10.4 *Before conducting work at a wetland in Buckskin Canyon, Mae Burnette pauses to listen at the head of the spring*

Watershed Program emphasized the need to learn directly from the land by caring for particular areas (see Figure 10.4). Elders have described these longstanding traditions as 'drinking from places' (Basso, 1996, pp85–86) and gaining 'a vision for the land' (Long et al, 2003). Through such direct experience, individuals can cultivate and demonstrate the proper frame of mind to apply traditional knowledge and new technologies in solving novel problems.

Honor learning that values tradition and collaboration

When outside education becomes associated with status, it often competes with traditional ecological knowledge, rather than adding to it. Statements and policies that afford special status to community members with college degrees can be seen as devaluing those who do not have degrees. Declarations or mere perceptions that employees who complete degrees will be first in line for promotions and pay increases reinforce the belief that education is an undertaking for personal, not community, advancement. An emphasis on personal achievement conflicts with an Apache norm emphasizing humility. Individuals who declare that they have particular kinds of knowledge may be considered boastful and disrespectful, and therefore likely to lose that knowledge or suffer some sort of spiritual retribution. In addition, there is a concern that knowledge that is made freely available may be misused – in extreme cases, for greedy or malevolent purposes. For example, practitioners of traditional herbal medicine fear that their knowledge could be sold or used for purposes other than healing. Consequently, traditional attitudes toward knowledge run counter to the norms and expectations of academic research, which emphasizes the validation of one's knowledge through publication.[3]

Several staff members of the Watershed Program emphasized the importance of humility and teamwork in completing projects, which they contrasted with the individual achievement represented by a college degree. For this reason, they argue that new employees need to prove themselves by completing projects that involve manual labor and working closely with local land users and other community members. For issues involving land, in particular, individuals with recent college degrees may be seen as lacking authority because they tend to be young and inexperienced. Young graduates risk violating tribal social norms by assuming that a college degree, rather than wisdom and experience, confers authority to make decisions.

Participatory research methods can help young students in natural resources to learn to ask and answer questions through collaborative work. Participants teach each other while avoiding the lecture styles and authority relationships that may generate friction. Changing the patterns of communication and teaching is critical for social learning and organizational growth. For example, informal and non-formal training methods, such as role-playing and group problem solving, are often more effective than conventional lecture-based teaching (Uphoff, 1986;

Stringer, 1997). We found that staff members learned more through group exercises and field activities than through lectures. The process of designing and conducting participatory research also helps students to learn how to communicate well, to address conflicts and to interpret how social networks and organization structures affect decision-making.

Communicating to promote trust and shared understanding

Participatory research methods often teach that leadership is less about making decisions than about improving communications (Grundy, 1996). We found that conflicts were often rooted in failures to communicate early and well. Trust must be built before participatory research can begin and communication must remain constant thereafter (O'Fallon and Dearry, 2002, pp157–158).

Bilingual ability, in both speech and thought, is an important asset for community problem-solving (Adley-Santamaria, 1997, p141). Fluency in an outside language helps community members to leverage support from beyond the community, while speaking the local language helps to establish trust within it. Fluency in the Apache language is often associated with the ability to provide rich descriptions or 'vision' – the ability to perceive and describe sites and their potential. For example, tribal council member Lafe Altaha praised a staff member's presentation on spring conservation: 'This presentation was really good. It was all done in Apache', and then added: 'I could really picture those places in my mind.' Visual methods work particularly well for describing places to elders, as Mae Burnette explained: 'Elders like to see those pictures to see what it looks like now – because they only saw it when they were small on horseback.' She added that describing places visually helped to avert resistance to management actions: 'Most of the elders don't have a problem with it. We put a picture in their mind of how it's going to look.' People who can 'speak from the heart' demonstrate that they believe in the value of a proposed project. Such demonstrations help community members to trust an individual, which, in turn, helps to ensure community support for a particular intervention (Long et al, 2003).

Because the Apache language has traditionally been transmitted orally, writing imposes barriers to shared understandings. Yet, proposals and reports written in English are the standard currency for most outside sponsors of research. Because such documents are difficult to bring into active discourse among community members, they are less helpful in promoting local understanding of a project. Participatory research projects have gravitated towards visual media as a means of promoting community participation (Odutola, 2003). We found that visual techniques such as poster displays, repeat photography, digital video and maps were effective in describing ecological changes. Videos, like the one being filmed in Figure 10.1, have allowed people to experience the vitality of the land through their eyes, their ears and their native tongue. Although these devices can speak differently to viewers from different backgrounds, they avoid much of the need for translation between languages.

Interpreting and managing conflict and power relationships

Effective researchers and managers also need skills in interpreting and managing interpersonal and social situations. Coordinating a participatory research project can provide experience in these skills, which are not generally taught in university classes on research methods. Individuals working in science-based fields often narrowly define their work to ignore these interactions. However, conflicts frequently arise in which people of different status end up blaming each other for 'not doing their jobs', rather than trying to understand the social basis of their conflicts (Putnam, 1996). When such tensions persist, they can easily undermine management efforts. In small communities, in general, decisions are often attributed to the 'politics' of envy and greed; but it is often unclear whether those purported interests lie at the level of an individual, a family or a larger group. The social orientation of participatory research may lead the researchers to examine the broader arena in which decisions are made within the community or an organization.

Tribal and non-tribal members often face different constraints when making decisions. When tribal member managers propose changes in the organization, their motives may be questioned because they may be seen as having more to gain from it. In addition to this scrutiny, tribal members are often expected to understand and adhere to cultural norms and traditions. They may be looked down upon for not being able to speak the native language. Outsiders are more easily excused from observing those norms and upholding those traditions. For these kinds of reasons, tribal members often face greater resistance among employees than do non-members when appointed to a management position within a hierarchical system of 'bosses and workers' (Trosper, 1988). These dynamics can create opportunities for non-members or outsiders to promote particular policies since they can bear a shield of neutrality. However, researchers can rapidly lose the appearance of neutrality and the advantages that it confers, particularly as they cultivate strong working relationships with specific tribal members or groups.

By encouraging community members to become agents of change, action-oriented participatory research becomes an inherently political endeavor. Researchers with backgrounds in the natural sciences typically have not been trained to understand the political and historical dimensions of their work. As a result, they often regard decision processes as a confusing, perhaps even insidious, black box of politics and they may fail to recognize how their actions can have unintended consequences. For example, a grant to conduct research activities or provide training positions for tribal members can create incentives to assert control over those new resources. Such competition can waste time and money, as well as create conflicts among community members. These problems can be a legacy of research that survives long after the researchers have left the scene. Research activities need to avoid creating new classes of elites who dominate resources and knowledge.

The procedural requirements common to participatory projects, such as obtaining permissions, arranging compensation for community members and determining how results will be used, help researchers to more fully consider the political and ethical ramifications of their work. Even more importantly, the process of collaborating with community members helps to understand and manage the social, political and ethical implications of a research project. Participatory research approaches seek to avoid concentrating authority and information within individuals by having community groups assign job responsibilities, by discouraging specialization and by rotating people through positions (Uphoff, 1986). Dispersing knowledge among individuals reduces the potential for any one individual to monopolize knowledge or drain it from the institutions should they leave (Elden and Levin, 1991).

Unfortunately, it is difficult to involve many community members in research efforts because there are few people with time and interest, and there are limited resources to support them. In addition, local institutions may not be well suited to promoting broad community participation. Many community institutions have evolved hierarchical structures to present a unified voice to outsiders, rather than facilitating broad participation in decision-making (Cornell and Kalt, 1995). In these situations, it is critical to allow time for more members to become involved and for local institutions to become more participatory.

CONCLUSION

To solve ecological challenges, communities need to cultivate problem solvers who can maintain an active and enduring interchange between inside and outside knowledge systems. Unfortunately, many people who leave their home communities to attend university become isolated rather than prepared to help solve problems within the social and cultural setting of their community. Universities and other institutions may contribute to this problem by conferring special status upon students for their technical knowledge, rather than their ability to conduct, share and apply research ethically and wisely. Neither 'technology transfer' to communities nor conventional education of community members adequately cultivates shared understandings across cultures. While transferred technology often sits idle, college-educated staff members who lack experience in addressing problems at the community level can become frustrated by their inability to effect positive change.

Through participatory research, students can gain experience in working with members of their community to solve problems. It helps them to consider the ethical implications of their work; the social and historical setting in which decisions have been made; and strategies for improving communication, managing conflict and engaging a broad spectrum of the community. These skills can promote a local knowledge system that evolves deliberately in response to new challenges and new information. Adapting local systems requires time and patience, as 'successes' collapse and 'failures' emerge as successes (Uphoff, 1986). As one elder advised her grandson, a restoration project manager (and one of the authors: Delbin Endfield): 'Go slowly. Listen to the land and it will tell you what to do.' People dedicated to fostering community research in ecological management should heed this advice in order to ensure that their efforts to erect bridges do not disrupt community systems that are already working.

Returning to the bridge

By proposing to clear the stream underneath the bridge in Cibecue without understanding the history of that place, the emergency rehabilitation team perpetuated the outsiders' tradition of dismissing the traditional values and knowledge of the community as outmoded. Some community members countered that they could live without the bridge, but not without the stream underneath. However, by contending that it would be better for the bridge to be washed out than to sacrifice the streamside growth underneath it, those community members were undervaluing their connection to the outside world, which provides vital services such as emergency healthcare. They also seemed to discount the stream's capacity to regenerate itself following the anticipated floods. Neither side fully acknowledged the risks of different responses; consequently, each was prone to making a poor decision. In the end, a compromise was reached, allowing the team to remove debris, but constraining them to an area only 9m (30 feet) above and 9m (30 feet) below the bridge. Some community members felt that the bulldozer operator did not follow those restrictions closely enough; but they were happier about the outcome than what happened at the second bridge in town. At that location more extensive dozer work was performed and significant bank erosion subsequently occurred (see Figure 10.5).

Local problem-solvers continued to watch the bridges to see how the stream responded so that next time they would find better answers. Alternatives that they have considered for those situations include relocating the bridge to a site where it would be less likely to affect the stream and to design fords or low water crossings over which the stream can flow. Author Mae Burnette reflected on what had happened at the bridge:

> *I'm pretty sure that where the bridge is now, there was a water crossing before. So the people there could get by if the bridge was gone. They [agency officials] spend all this money to get the young people an easy way out. If Mother Nature doesn't like what has been done, it will get rid of that bridge. The people will adapt; that's what they've always done.*

Source: Jonathan Long (July 2003)

Figure 10.5 *Staffers from the Watershed Program examine bank erosion following clearing of debris at the lower bridge in Cibecue*

ACKNOWLEDGMENTS

We wish to thank our fellow researchers in the Tribal Watershed Program, including Floyd Cheney, Michael Cromwell, Arnold Pailzote and Daryl Tenijieth, who contributed their ideas on developing ecological management on the reservation, as well as the residents of Cibecue, who have valiantly defended their land.

NOTES

1 An early draft of this chapter was published in 2004 by Utah State University in the *Proceedings of the Fifth Biennial Conference on University Education in Natural Resources* (Long et al, 2004).

2 The Bureau of Indian Affairs is the branch of the federal government responsible for the administration of reservation lands held in trust by the US for American Indian tribes.

3 We have not explored the issue of community held versus individually held intellectual property in this chapter, which Hankins and Ross (see Chapter 11 in this volume) discuss in depth. Most of our work sought to apply and share knowledge within the

reservation, rather than to publish the knowledge in academic outlets. There are strong informal sanctions against taking credit for such knowledge as one's own. However, there are other types of knowledge (e.g. stories, songs and uses of plants) that are often considered to belong to individuals.

REFERENCES

Adley-SantaMaria, B. (1997) 'White Mountain Apache language: Issues in language shift, textbook development and native speaker-university collaboration', in J. Reyhner (ed) *Teaching Indigenous Languages*, Northern Arizona University, Flagstaff, AZ

Basso, K. H. (1996) *Wisdom Sits in Places: Landscape and Language among the Western Apache*, University of New Mexico Press, Albuquerque, NM

Buskirk, W. (1986) *The Western Apache: Living with the Land before 1950*, University of Oklahoma, Norman, OK

Cornell, S. and Kalt, J. (1995) 'Where does economic development really come from? Constitutional rule among the contemporary Sioux and Apache', *Economic Inquiry*, vol 33, pp402–426

Dalton, R. (2004) 'When two tribes go to war', *News@Nature* 430(6999), pp500–502, www.nature.com/news/2004/040726//pf/430500a_pf.html, accessed 7 July 2006

Elden, M. and Levin, M. (1991) 'Co-generative learning', in W. F. Whyte (ed) *Participatory Action Research*, SAGE Publications, Newbury Park, CA

Grundy, S. (1996) 'Towards empowering leadership: The importance of imagining', in S. Toulmin and B. Gustavsen (eds) *Beyond Theory: Changing Organizations through Participation*, John Benjamins Publishing Company, Philadelphia, PA

Hiller, J. G. (2005) 'Is 10% good enough? Cooperative extension work in Indian country', *Journal of Extension*, vol 43, no 6, www.joe.org/joe/2005december/a2.shtml, accessed 27 August 2007

Levin, M. (1999) 'Action research paradigms', in D. J. Greenwood (ed) *Action Research: From Practice to Writing in an International Action Research Development Program*, John Benjamins Publishing Company, Philadelphia, PA

Long, J. W. (1994) *Building Connections: A Strategy for Integrating Natural Resource Management*, MPP thesis, Harvard University, Cambridge, MA

Long, J. W. (2000) 'Cibecue watershed projects: then, now and in the future', in P. F. Ffolliott, M. B. Baker Jr, C. B. Edminster, M. C. Dillon and K. L. Mora (eds) *Proceedings of Land Stewardship in the 21st Century: The Contributions of Watershed Management*, 13–16 March 2000, Tucson, AZ, RMRS-P-13, US Forest Service Rocky Mountain Research Station, Fort Collins, CO, pp227-233

Long, J., Tecle, A. and Burnette, B. M. (2003) 'Cultural foundations for ecological restoration on the White Mountain Apache Reservation', *Ecology and Society*, vol 8, no 1, p4, www.ecologyandsociety.org/vol8/iss1/art4/, accessed 26 March 2004

Long, J., Endfield, D., Lupe, C. and Burnette, B. M. (2004) In T. E. Kolb (compiler) *Proceedings of the Fifth Biennial Conference on University Education in Natural Resources, Natural Resources and Environmental Issues*, vol XII, Logan, UT, Quinney Library, College of Natural Resources, Utah State University, www.cnr.usu.edu/uenr/nau/UENR5_proceedings.pdf, accessed 19 March 2008, pp29–44

McDonald, D. A., Peterson, D. J. and Betts, S. A. (2005) 'More tips: What if a cooperative extension professional must work with Native American institutional review boards?', *Journal of Extension,* vol 43, no 5, www.joe.org/joe/2005october/tt1.shtml, accessed 29 August 2007

Medina, A. L. and Long, J. W. (2004) 'Placing riffle formations to restore stream functions in a wet meadow', *Ecological Restoration*, vol 22, no 2, pp120–125

Mehari, A. (2007) *A Tradition in Transition: Water Management Reforms and Indigenous Spate Irrigation Systems in Eritrea*, Taylor and Francis Ltd, London

Niamar-Fuller, M. (1998) 'The resilience of pastoral herding in Sahelian Africa' in F. Berkes and C. Folke (eds) *Linking Social and Ecological Systems: Management Practices and Social Mechanisms for Building Resilience*, Cambridge University Press, New York, NY

Odutola, K. A. (2003) 'Participatory use of video: A case study of community involvement in story construction', *Global Media Journal*, vol 2, no 2, art 11, http://lass.calumet.purdue.edu/cca/gmj/sp03/graduatesp03/gmj-sp03grad-kole.htm, accessed 9 July 2006

O'Fallon, L. R. and Dearry, A. (2002) 'Community-based participatory research as a tool to advance environmental health sciences', *Environmental Health Perspectives Supplements*, vol 110, no S2, pp155–159

Petras, E. M. and Porpora, D. V. (1993) 'Participatory research: Three models and an analysis', *American Sociologist*, vol 24, pp107–125

Phillips, L. (2006) 'Communicating social scientific knowledge dialogically: Participatory approaches to communication analysis and practice', in N. Carpentier, P. Pruulmann-Vengerfeldt, K. Nordenstreng, M. Hartmann, P. Vihalemm and B. Cammaerts (eds) *Researching Media, Democracy and Participation: The Intellectual Work of the 2006 European Media and Communication Doctoral Summer School*, Tartu University Press, Tartu, Estonia

Putnam, R. W. (1996) 'Creating reflective dialogue', in S. Toulmin and B. Gustavsen (eds) *Beyond Theory: Changing Organizations through Participation*, John Benjamins Publishing Company, Philadelphia, PA

Ruano, S. (1991) 'The role of the social scientist in participatory action research', in W. F. Whyte (ed) *Participatory Action Research*, SAGE Publications, Newbury Park, CA

Santos, J. L. G. (1991) 'Participatory action research: A view from FAGOR' in W. F. Whyte (ed) *Participatory Action Research*, SAGE Publications, Newbury Park, CA

Stringer, E. (1997) 'Teaching community-based ethnography' in E. Stringer (ed) *Community-based Ethnography: Breaking Traditional Boundaries of Research, Teaching and Learning*, Lawrence Erlbaum Associates, Mahwah, NJ

Taylor-Ide, D. and Taylor, C. E. (2002) *Just and Lasting Change: When Communities Own Their Futures*, The Johns Hopkins University Press in association with Future Generations, Baltimore, MD

Trosper, R. L. (1988) 'Multicriterion decision-making in a tribal context', *Policy Studies Journal*, vol 16, no 4, pp826–842

Uphoff, N. (1986) *Local Institutional Development: An Analytical Sourcebook with Cases*, Kumarian Press, West Hartford, CT

Research on Native Terms: Navigation and Participation Issues for Native Scholars in Community Research

Don L. Hankins and Jacquelyn Ross

Introduction

> *Field biologists promoted California's first clam harvest regulations midway through the Great Depression, [nearly three-quarters] of a century ago. Their purpose was environmental protection. By limiting the number of clams that could be harvested, they sought to conserve the clambeds of areas like Tomales Bay in Marin County. [At present], Tomales Bay has yet to recover from the good intentions. Of the three clam varieties that crowded its beaches in 1935, two were extinct by 1945 (including a prized but tough horseneck) and the third was barely hanging on.*
>
> *What had caused the death of the Tomales Bay clambeds? The Coast Miwoks of the area were among those who opposed the 1930s regulations that severely limited the harvest. It was the act of harvesting, they insisted, that was keeping the clambeds healthy.*
>
> *In 1980, state fish and game biologist Walt Dahlstrom reviewed the situation and confirmed the view of the Coast Miwoks. 'Well, steady cultivation of clam beds definitely improves them... When they stopped digging the way they used to, there was really a good bit of loss because the young clams had no room to grow.* (Baker, 1992, pp28–29)

As this brief history of clam bed management in Tomales Bay shows, traditional academic research has largely excluded or ignored native community concerns and needs. At best, the involvement has been primarily extractive, where information, local contacts and other forms of assistance are provided by a host community, with little or no reciprocity on the university side. At worst, damaging policy decisions are made on the basis of research that is conducted without community understanding, guidance, involvement or informed consent. Tuhiwai Smith (1999, p2) addresses the notion that research inevitably benefits people: 'It becomes so taken for granted that many researchers simply assume that they as individuals embody this ideal and are natural representatives of it when they work with other communities.' Indigenous communities in the US and elsewhere have different experiences. The case of the clam beds at Tomales Bay is an example of research leading to regulations that ignore traditional cultural management and damage a tribal subsistence economy. This case and others like it have led many native communities to be extremely wary of formal research and other external processes.

Conducting research in and for native communities thus involves challenges that some researchers may find unique or unexpected. Competing expectations from, and accountability to, the host community, faculty, funding agencies and other parties can pull researchers in opposing directions (see Chapter 3 in this volume). The problem is especially acute in native communities because of the legacy of extractive research.

In such an environment, it takes deliberate preparation and communication to reach understanding and agreement with host communities about appropriate research questions, protocol and process. How do native people explore, apply and/or share unique community-held traditional knowledge in the contemporary research setting without exploiting or being exploited? Participatory research (PR) is an approach that can accomplish research needs for the benefit of the community-focused scholar and the community itself. It can provide communities and scholars both with new tools and foster community access to, and interest in, the resources of the university. Should this lead to communities encouraging the development of their own scholars, there is powerful potential for useful research in the natural resources. Yet, effectively engaging in a PR process requires an appreciation of some issues that are commonly encountered by native scholars. Engagement also requires clear communication strategies between native communities and researchers. When strong relationships based on good communication are achieved, there is potential for mutual expansion of native and scientific knowledge.

In our experience, effective PR is research conducted within a community that addresses questions and topics of interest to the community. Such research is generally guided, conducted, engaged in and approved by the community members. The community helps to determine how best to collect data, interpret results and share those results in ways that are not only accessible to the community, but also to academic outlets, which are often the controllers of such information from native communities. Achieving such high levels of community involvement requires sustained effort in establishing and maintaining relationships of trust, respect and reciprocity (see Chapter 5 in this volume for an example of such an effort).

ISSUES FOR THE NATIVE SCHOLAR

The need for respect is acute in any PR project and especially so when collaborating with native communities because of the legacy of scientific researchers ignoring and dismissing native systems of knowledge. The more recent quest by scholars for traditional ecological knowledge (TEK) has raised ethical issues about information ownership and sharing, exploitation and the implications of using only select pieces of knowledge systems that function holistically. Conventional research practices continue to create multiple concerns as researchers and scholars inside and outside academic circles consider how to, or even whether to, participate in research or other processes. These issues include:

- whether the research will be viewed objectively versus subjectively;
- the danger of becoming isolated by colleagues because one's research is considered 'ethnic' and therefore not connected to the mainstream;
- managing differences in worldviews and philosophies;
- acknowledging and contending with centric perspectives and history in academia;
- understanding tense and the relevance of temporal scale;
- walking in multiple worlds; and
- being on view.

Objectivity versus subjectivity

With any research project, it is important for a project team – the scholars and host community members – to consider the objective and subjective nature of the project. Does the research question have value for the host community? Is the project designed for empirical data collection with specified criteria where the data collection and interpretation is unbiased, or is it designed to collect only the data that lends support to a specific viewpoint or cause (i.e. there is significant researcher/community bias in data collection and interpretation)? In the context of PR, scholars must consider how their involvement in a research project is related to the role of the scholar as a researcher versus an advocate.

This issue is particularly of concern for native people researching native subjects, or subjects pertinent to native communities. Within the community, the research question and the objectivity/subjectivity issue is particularly critical because the research results may invalidate information of practical importance. For instance, if a host community acts on the finding that traditional land management practices are no longer compatible to sustainable yield of a desired plant species, this could result in the loss of responsibility or status among the community members who guide those practices. What does it mean to the community if the research results do not support cultural practice and tradition? How can the community ensure that their researchers have what is needed to consider and gather information in an appropriate cultural context: one that reflects the community framework? Certainly, the intent to conduct PR is to address the pragmatic concerns or needs of the community. Thus, establishing an objective research framework can assist in developing new research questions and options in such if/then situations.

In any research project, scholars must guard against bias and must be especially vigilant when conducting research intended to benefit a community. Deliberate bias is unethical, and in PR projects, unintentional bias can be warded off by a discussion between scholars and host communities about the definition of research bias, how it can compromise results, and the potential vulnerabilities for bias in their particular research venture. It is important to establish high personal ethical standards, and in discussing these standards openly with host communities, scholars can enlist community help to stay on course. With such disclosure, potential misunderstandings can be mitigated and the scholar and the native community can serve as a role model for others within the context of the scholar–community partnership to uphold such ethical standards.

There is often the assumption that a researcher studying something close to heart will bias the results in support of their beliefs or in deference to community pressure for a desired result. Due to this assumption, research from scholars who come from communities who are not well known or understood in the academic world may be scrutinized to a greater extent than would research from scholars external to the community. Thus, how the research is conducted, received and evaluated has bearing both for the scholar and community involved. Such considerations apply to research across disciplinary boundaries (e.g. social sciences or physical sciences). When the scholar is from or associated with the host community, both the scholar and the community are 'the other'. What mainstream scholarship perhaps does not consider enough are the high standards of 'the other' and the pragmatic roots of such standards. In a discussion about the practicality of Native American cultures (Bosk, 2004, p11), ecologist Dennis Martinez is quoted as saying: 'Lies would lead ultimately to starvation: you report accurately what you see, what you don't see.' Scholars should bear in mind that in PR, host communities will scrutinize and evaluate the research and use their own measurements *à propos* to the local setting. In discussing the community impact of academic programs and community development projects, Boyer (2006, p15) notes that a community may decide a project is worthwhile, even if it is not succeeding by Western standards. Expectations and possibilities need to be discussed at the outset of a project.

The 'ethnic' categorization

Due to the rarity of native scholars and, in many cases, applied research among native communities, native scholars and/or research among native communities can often be stereotyped or classified as ethnic – an ethnic 'otherness', so to speak. Ethnic categorization has served a major role in academic and public institutions. For instance, equal opportunity programs have acknowledged the lack of access to the academic world and have provided openings to native researchers where these opportunities may not have existed previously. There may be funding available to assist scholars from under-represented communities. One potential shortcoming of such categorization, however, is the labeling of a scholar or one's research as 'ethnic' simply because of the color or status of the researcher and/or the community. Such categorization can be a great disservice, particularly where the research itself does not involve a question or topic related to ethnicity. For instance, ecological studies investigating traditional native land management practices might be labeled as ethno-ecological or ethno-botanical even though the primary focus of the research is ecological in nature. In such cases, the research and important connections to other researchers and projects may be overlooked due to the ethnic labeling. The research should be valued for the scientific content. From the perspective of community utility, it is more relevant to cite good research regardless of the ethnic background of the individual. Within the ethnic otherness there is also the possibility of automatic association with others in the ethnic category, regardless of any genuine research relevance. For example, the recommendation that 'You should see X's paper, they worked with natives', is not particularly useful if there is no similarity of research theme or problem. This situation can be especially problematic when scholars know that although X or X's project is much lauded in the university world, it is poorly regarded in the native world.

Differing worldviews and philosophies

There is a natural conflict that exists between native worldviews and those of the academic world and Western society at large. In a general sense, native worldviews are holistic, encompassing the physical and metaphysical interactions of life. For instance, a native worldview may not categorize distinct boundaries between humans and nature; rather, humans are one component of nature. Furthermore, natural elements and features such as water, soil and rocks may be considered animate, and may be analyzed in terms of the relationship of everything else in the environment, the history and the pre-history of the community.

In contrast, the Western research world draws distinct boundaries between these categories. Specifically, the scientific method depends upon breaking things down and analyzing them in their separate (i.e. disconnected) components. While some disciplines (e.g. geography and ecology) enable a merging of methods and worldviews, there is little interaction between categories of nature and/or the human–nature relationship. Water, soils and rocks are not viewed to be animate, although they demonstrably support life and have their own energies. Such philosophical differences can pose problems in a research context, particularly where causal relationships are being investigated.

At present, it is imperative to discuss such concepts or causal relationships in a common language framework, and such a framework may be reached through discussion between scholars and host communities. For community members accustomed to tradition that connects everything and everybody to a larger story, the specialization and compartmentalization inherent in conventional research can seem absolutely foreign. This does not mean that the community would reject such practices – especially in light of a research question's urgency, or available resources; but an acceptable framework and mutually understood terminology for the research should be discussed and understood (see Chapter 10 in this volume for a discussion of integrating indigenous and scientific knowledge). This discussion may well build the capacity of scholars and the community members alike, opening the door to more collaborative interdisciplinary investigation.

Acknowledging centric perspectives in the academic world

Academics often recognize and pay homage to the 'mothers and fathers' of their specific disciplines. There is an adherence to principles that are based on landmark studies in the field until the principles are disproved.

It is important for native scholars to acknowledge the conventional academic lineage as it can lead to a discussion of how the field has evolved over time, and helps to illuminate the differences in perspectives between native communities and academic communities. Developing an understanding of the cultural differences between native communities and academic communities can also help to draw important parallels, especially when Western academic principles and traditional cultural principles align. The conventional history of academic disciplines and related research can be helpful and, in some cases, even analogous to the way in which oral history is shared in the tribal context. For instance, parallels between Darwinian evolution and similar evolutionary discussions encoded in creation stories can be drawn to find common ground, so to speak, between academic and native community principles of thought.

This common ground is rarely acknowledged, however. What might seem especially strange to the native scholar is the comparatively short historic timeline of the academic field, especially in cases where the line of native enquiry in the same direction is far longer. The disparity – and the feeling of waiting for the world to catch up to time-honored, stable knowledge – finds deft expression in Osage writer Duane BigEagle's poem 'My Grandfather Was a Quantum Physicist'. He describes his grandfather in a traditional ceremonial dance setting, then notes that scientists are now looking 'beyond the stars/ and beyond time' to find the things that his grandfather already knew.

While learning about the timeline of developments in their academic fields, and adjusting to a system in which something that is absolutely true today may be rejected in the near future, native students may quickly realize that some fields were born among a research community that had little diversity of any kind. This can be alienating because of the lack of diverse perspectives and/or the recognition of the contributions of native communities to the disciplines. Giving credit, then, to the academic lineage without acknowledging the pre-existing work of cultures that primarily utilize an oral tradition might become problematic. The situation can become even more complicated when an academic scholar is given credit as an individual for unearthing and then publishing information that is community-held knowledge of a tribal people. A similar problem occurs when a new generation of scholars cites and builds upon previous work that is thought to be erroneous or invasive. This may be viewed as theft and/or betrayal.

Understanding tense and temporal scale: Bringing the academic world into the present

One of the more persistent problems encountered by the scholar working with native communities is the academic (and public) perception that there is a huge chasm between past and present knowledge in the native community; that, in fact, many non-natives view today's native people as different and disconnected from their ancestors. Traditional tribal basket weavers in California found this out when they banded together in 1992 to form the California Indian Basketweavers Association (CIBA). Weavers and CIBA staff started meeting with state and federal agency officials to discuss access and health issues connected to basketry, only to find that the officials were dumbfounded to learn that tribal people still gathered natural plant materials for weaving and for food. The policy implications of this lack of knowledge were serious, for weavers and food gatherers were harvesting in traditional gathering areas managed by agencies and private companies for silviculture, agriculture and road maintenance. This management often involved the application of herbicides and pesticides, without consideration for possible human consumption. Training and communicating with agency officials and scientists have highlighted such concerns and have become an important part of CIBA's work (CIBA, 2007).

There is often an assumption that Native communities have not retained ancient environmental knowledge; however, many have, and others are actively involved in revitalizing such knowledge. Another common manifestation of this assumption is the use of past tense when referring to Native life ways, knowledge and skills. Good examples of this tendency towards using the past tense can be seen in written interpretive material in such venues as state and national parks in the US, as well as public school textbooks. For instance, it is common to visit national parks and see interpretive panels stating that *the Native Americans of this region managed the resources*, even though such practices still occur (albeit perhaps at a lesser scale). It can be a challenge for the scholars and native community members to bring mentors, teachers and colleagues forward into the present. If contemporary research is to be taken seriously, the persistence and continuity of native knowledge must be recognized.

Walking in (at least) two worlds

Not all native scholars are bicultural (i.e. simultaneously practicing Western and their own native traditions) when they start their research. Some may never be. It is not safe to think that all native scholars have a 'traditional' upbringing, and the definition of just what that is may vary widely according to the scholar's personal and community history. For students who do come into research with a strong cultural grounding, they may find that such a foundation gives them great personal strength, confidence and a framework for thought that allows for risk taking and creativity. Several years ago, the authors led a workshop on PR at the American Indian Science and Engineering Society (AISES) annual conference in New Mexico. AISES is an organization that promotes the representation of American Indian and Alaskan Natives in engineering, science and related technology disciplines (AISES, 2008). The organization advocates very high levels of scholarship and intensive career exploration. Although our workshop was aimed at high school (secondary) students, participants came from a variety of ages, professions and grade levels. After we had a lively discussion about some of the challenges of working on indigenous issues in largely non-indigenous institutions and companies, we asked the high school students what they thought they brought to their prospective institutions in addition to stellar academic records and a desire to learn. The answers were fascinating, and are summarized in Box 11.1.

Box 11.1 Native scholar's toolkit

What do Native American students think that they will bring with them to college? Here is a short summary of the tools that a group of American Indian Science and Engineering Society (AISES) students identified as culturally based skills and attributes.

Language skills

Being fluent or learning a tribal language helps students to be more flexible in their thinking and offers different (often holistic) reference points that are not present in English.

Listening

Learning how to listen is a skill that may be a product of largely oral cultures. Students mentioned being taught to sit still and listen for long periods of time as young children. This helps their attention and concentration in school.

Patience

Having patience and not rushing enables students to adhere to tasks and process in their schoolwork.

Respect

Having 'built-in' respect for elders and other knowledgeable people often translates into respect for school faculty, as long as the faculty also respect the students. Being in an academic environment that encourages argumentation or even conflict can take some adjustment, and students expressed concern about how to disagree without being disrespectful or risking personal attack.

Relationships

Sensitivity to the complex, intertwined relationships within tribal communities was seen as a benefit, and may make students more invested in discussion and consensus towards resolution when problems arise.

Religious (spiritual) beliefs

Deeply rooted belief systems give students strength, confidence, stability within the academic environment, and provide connection to the home community.

Knowing protocols

Acknowledgment and acceptance of long-established cultural protocol was seen as a benefit. Learning the complex protocol and hierarchy of academic life was not seen as being particularly foreign, unexpected or daunting.

Leadership

The question 'What leadership skills are important in the tribal world?' yielded the following skills and characteristics:

- *Community representation:* recognizing that the community has chosen a person for leadership means that the accountability must be directly to the community as a whole, and to each person as an individual. Good community representation means that there must be extensive consultation and discussion before decisions are made.
- *Attitude:* a 'good' attitude should be one of looking for the positive, viewing oneself as accountable to others and being willing to learn.
- *An awareness of 'Indian time' and chronological order:* this was explained as doing things when the conditions 'are right' and not being dictated by a strictly predetermined schedule. Students spoke to the issue of sensitivity to the rhythms within the community, 'allowing appropriate time for things to work out as needed ... don't shoe-horn things'. Also mentioned was the principle that the cultural/spiritual observance of life events may supersede secular activities.
- *Knowing your people:* students addressed the importance of the deep knowledge of a community being essential to providing good service to the people.
- *Knowing oneself:* being aware of one's own strengths, vulnerabilities and fears is a form of honesty that was highly valued by students. An assessment of such aspects was seen as key to planning for improvement and education.

Even exceptionally equipped students find the academic life to be unique. Most native scholars, and even first-generation scholars of Western backgrounds, need to learn the culture of the academy. It is here that faculty members can be of tremendous assistance, guiding their students through the protocol, teaching the language, strategies and behavior that will help scholars balance their academic life. Learning more about the skills and attributes that new scholars bring with them will help faculty to gauge how and where to guide their students. As in any learning situation, good communication – especially good listening – on both sides of the relationship will help.

You are always on view

The 2000 US Census suggests that Native Americans and Alaska Natives comprise approximately 0.9 per cent of the US population. Typically, within academia, native people are disproportionately underrepresented. Thus, native researchers are quite unique among academic circles. Due to the historic associations between the dominant culture and native cultures, it is important to recognize that native researchers and native communities are always on view. In many ways, native individuals are the diplomats between the outside world and tribal communities. Thus, it is important to keep this in mind throughout one's pursuits in research or the academy.

Similarly, native people are frequently associated with the popular media depiction of natives people and issues. Furthermore, native individuals are often stereotyped by public perception about native issues. For instance, there is often an association that native people are environmentally astute or, more recently, that native people have adequate or even excessive financial backing as a result of large casino enterprises. These misconceptions can even be perpetuated in the academic world. Certainly, there are plenty of examples of poor environmental conditions and extremely modest living conditions for many native communities. For scholars, it can be difficult to remain focused when there are so many detractors. However, one's research has the potential to shift the mode of thought generated by the dominant culture about native peoples. Thus, the work by native researchers might call for a higher standard in order to bring attention to the fact that there is diversity beyond the stereotypes.

CREATING BETTER RESEARCH PRACTICE

The relationship between the academic world and native communities is the central issue. Undoing the historic wrongs of extractive research and vanquishing the dilemmas of the native scholar rests on the development of better relationships and the acknowledgment of what has happened in the past. While many factors go into the development of mutually beneficial relationships in research, we suggest that three are particularly important:

1 reciprocity, or giving something in return;
2 developing pertinent research that is utilitarian or pragmatic; and
3 recognition of intellectual and community property rights.

PR is a labor-intensive methodology and even the most productive, positive ventures require cooperative giving from the host community. Time, information and energy could be spent elsewhere; so when community members sacrifice for a research project, such selflessness should be understood and appreciated.

Reciprocity

Native communities have been subjected to extractive research for too long. Over the last century (and earlier), researchers entered native communities to collect ethnographic materials, record languages and customs, and document a variety of attributes of 'their' chosen native communities. While some of this information has proven useful to native communities struggling to retain their cultural identity in the contemporary era, much of this research has caused a great rift between academic and native worlds due to the fact that the information has been largely inaccessible to native communities. The question of who ultimately 'owns' community knowledge and cultural items is thus a major issue. At the heart of the issue is trust and respect. The principle of reciprocity requires scholars to share the results, ask the community if there are research skills that community members would like to gain, and acknowledge the community for their participation in the research process. Furthermore, the scholar should make available any materials (published, archived or scanned) that have been derived from the community.

Many native communities view information and cultural items collected from the community as their own responsibility despite the disposition of such things. Of exactly what benefit would it be to a native community to share information or cultural resources with a scholar if that research or collection would not be accessible to that community? As Schroeder et al (2006) note, the historic collection of archaeological artifacts and human remains and the modern collection of genetic materials from tribal peoples are painful, persistent issues among native communities because the communities lose control of their own heritage and bodies, and because the uses to which these artifacts and materials are put are often viewed as desecration. The patenting of, and profit from, biomedicines derived from genetic material and human genomics is seen as particularly egregious and profane by some peoples. Consider also the extraction of funerary items from native communities. For well over a century, the funerary items and remains of Native Americans have been collected, studied and stored in academic institutions. This has been particularly unsettling to many native communities and has a huge impact upon the ability to respect and protect one's ancestors and land.

The violation of cultural properties embedded within the landscape is also a concern for many native communities. For instance, revealing the locations of sacred sites, which many researchers have done in their publications, has opened up opportunities for infringement of traditional laws by researchers, by vandals intent on desecration and by curiosity/spiritual seekers at these sites. Such actions are examples of the continual disrespect by non-community members (i.e. academics and land managers) for traditional native cultures that contribute to the distrust of researchers among many native communities. Developing reciprocal relations between researchers and community members does not erase the past, but helps to reverse these trends and permits all collaborators in the research to learn and to benefit.

Utility

Native communities require pragmatic research approaches. What incentive do native communities have to be involved in research that does not address some fundamental attribute of interest to that community? It is of little use to research something that is already known or something that does not have a practical application. There is a plethora of practical concerns in tribal America, including health and natural resource issues appropriate for interdisciplinary study. Successful research among native communities often involves collection and analysis of data necessary to improve upon some existing condition. For example, Hankins (2005) analyzed the effects of prescribed fire, which included traditional seasonal considerations, on riparian ecosystems. This was pragmatic because it provided data upon which the community and land managers could base resource management decisions. This research was also relevant to land managers and others interested in knowing the effects of such management actions in riparian ecosystems.

Involving native community participants in developing research questions can help to identify conditions of concern to the community. There may be instances where the community has specific research questions that they would like resolved. Although those questions may not be entirely aligned with the scholar's research questions, it is important to incorporate, to the extent practicable, the research questions of the community. Doing so nurtures a relationship of reciprocity with the community in addition to contributing to practical outcomes. Where the researcher has established research questions, the inclusion of other questions or concerns by community members can enrich the research process.

Staying flexible throughout the research process can provide rewards to both the researcher and the native community involved. For instance, as discussed in Chapter 5, the burning of the sedge bed within the Native American Tending and Gathering Garden required both researchers and community members to keep an open mind in order to determine species level effects. The reward to the researchers and community members is that the collected data contributed to the baseline knowledge for the species managed, and the abundance of the resources increased, which meant more weaving materials for community members.

Intellectual and community property rights, citation and attribution

Research involving Native subjects may involve traditional cultural knowledge that is not proprietary. For instance, environmental knowledge or traditional ecological knowledge may stem from the collective knowledge of place acquired through generations of communal observation. Although some individuals may be the keepers of such knowledge, the foundation of that knowledge is considered communal property among many native communities, special societies or family groups. Thus, the distribution of such knowledge and the citation of such knowledge can be difficult to delineate. This is a discussion topic of paramount importance for scholars and their host communities. Where appropriate, the researcher should attempt to outline the basis of such knowledge within any manuscripts in order to maintain proper attribution of the knowledge as community or family property. This is not to exclude attribution to any individual community member. Rather, it is a means of clarifying that the individual whom one may be citing is really the intermediary of knowledge between the community and the researcher and that she or he has community or family permission to act as such. This is important for the sanctity of the information and for the reputation of all involved. For instance, Anderson (2005) cites traditional cultural practitioners for their knowledge of various land management practices. The knowledge discussed in the text is often widely known among the communities from whom the information was gained. In some instances, it may be appropriate not to disclose the individuals who shared such knowledge.

STRATEGIES FOR COMMUNICATION IN PARTICIPATORY RESEARCH

Reciprocity, utilitarian research, and recognition of intellectual and community property rights are key to nurturing strong relationships between the academic world and native communities. Good communication is fundamental to succeeding in these objectives. Good communication is not necessarily easy to achieve in any relationship, however, and the legacy of extractive research increases the risk of miscommunication with native communities. We therefore suggest that strategies for collaborating with native communities on research include attention to fair practice, ensuring accountability, establishing a common language, integrating culture and science, and assessing one's own skills and vulnerabilities.

Fair practice (decision-making, partnership)

Fair practice in PR involves respectful joint decision-making. This requires working with community leaders within the parameters of the community's own protocols for managing relationships with people outside the community. Many communities have developed their own gate-keeping systems to determine whom to let in when it comes to research. These may take the shape of informal meetings with respected leaders in a community or require a presentation to a formal tribal institutional review committee. In working through these channels, one may find that good social skills are as useful as a solid science foundation when it comes to community-based projects. Novice researchers may need to seek training outside of their discipline-based curriculum in order to feel prepared for leading focus groups and discussions, conducting interviews or, in some cases, mediating. Seeking out community leadership for such responsibility may be a viable option.

Once relationships are initiated, community members and researchers should discuss all the major components of the proposed research, including how community members can build individual and tribal capacity through and perhaps beyond the life of the research project. There must be agreement that the central research question has practical utility. That is to say, the answer to the research question will be of interest and benefit to the host community. The community understands proposed methodologies, and if conducting community meetings, interviews or surveys of community members is integral to the research, the researchers clearly explain at the outset the frequency and level of needed participation.

Some type of advisory committee or liaison between the community and researcher is extremely helpful. The authors have also found it helpful for faculty mentors or colleagues to meet the community members. This can foster lasting links between the community and the academy, and can lessen the likelihood of the student researcher becoming a liaison between the two. Many native communities experience a perpetual turnover among contacts at public agencies. Such turnover is disruptive to native community relationships with such entities and related projects. Student researchers, who likely will move on upon completion of their studies, pose a similar problem to native communities. Thus, having a connection between faculty mentors and colleagues within the academy can minimize the turmoil generated by potential changes among scholars working within the community.

Accountability (standards and consequences)

How does one build accountability into a research project? A chain of command, communications plan and process for resolution of disagreements should be discussed. The researcher is accountable to both the institution and the host community. The relationship to one's faculty adviser and the institution should be fairly straightforward; the relationship to the community may be more nebulous. It is helpful to talk about the best ways of making decisions in order to achieve understanding and agreement while keeping the work moving. Communications can be assisted by agreeing on guidelines or policies for the project at the outset. This can be especially useful when the research involves multiple partners or agencies. The development of such parameters will help to answer questions about who can join the project; how the project fits into the organizational structure of committees, departments or agencies within the host community; and who has access to the research process or information while the work is being conducted. This is also the place to discuss cultural practices or rules pertinent to the research. Creating such instruments helps to troubleshoot possible problems and to surface issues of concern to the various parties involved. The history of other research projects may come up in planning discussions. If a community has been extensively studied, it is important to know the impacts of such work. If there were problems, these need to be addressed in order to avoid the same mistakes. How was previous research perceived, received and interpreted in the community?

The exit plan should be discussed. How do the community and partners benefit from the research; who owns the results; how will results be shared and with whom; and what happens after the researcher leaves? All of these areas should be discussed as a matter of courtesy as well as efficacy.

Looking for cultural parallels

Establishing a common language is an essential process in conducting any community-oriented research. In some instances, the relationship to academic terminology, jargon or concepts can be difficult to translate into a format that can be comprehended by participating community members. One means of establishing a common language is to draw culturally relevant parallels to the subject matter. For instance, the discussion of evolutionary differences and divergence could include comparison to evolutionary processes described in some traditional stories. Many scientific concepts are encrypted in traditional stories, which can be interpreted for the scientific correlation. For instance, the concepts of intermediate disturbance hypothesis and conservation were certainly well known among most Native California communities due to the knowledge base and cultural application of fire as a traditional land management tool. The traditional knowledge base and stories within these communities have established a baseline of this sort of knowledge. Traditional cultural knowledge can also serve as the means to inform the research or academy.

Personal skills and vulnerability assessment

For new researchers, it may be appropriate to assess one's skills. 'Do I have the appropriate skills to conduct the research that this community wants? If not, what could I do to gain those skills?' Being realistic with one's abilities is critical to successful implementation of any research project. If the proper skill set is not in place, then the objectives will be difficult to achieve. In some instances, it may be appropriate to engage help from others more skilled, such as mentors, academic peers, community members and elders. Often, these individuals may have a sense of what one may need to be successful. Generally, the success of the 'mentee' is a success for the mentor too. Once areas of deficiency have been identified, attributes of the project can be developed to help implement useful proficiency.

Similar to assessing one's skills, the familiarity with one's weakness(es) is also an asset to being a successful researcher. If the weakness(es) can be identified prior to entering the field, then the researcher will minimize any self-inflicted stumbling blocks with respect to a variety of potential issues, which might include confusion between the role of researcher and community member. It is easy to become complacent about one's status if there are multiple identities involved (i.e. tribal member, researcher, a relative with vested interest in community). As a researcher, it is imperative to retain the mindset of a researcher while drawing on one's strengths and experience as a community member. While being an advocate for one's research is possible, it is more powerful for the host community to speak in support of the work and to let the research speak for itself.

In addition to knowing one's vulnerabilities, it is also valuable to know the vulnerabilities of the research team (i.e. the partnership of researcher(s) and community members). Assessing areas for improvement can facilitate learning and development within the group, and can increase the success of the project if project implementation pairs the strengths of team members with the needs of others. By doing so, the entire research team will gain from the research experience.

CROSS-POLLINATION AND ENRICHMENT

Developing strong relationships between the academy and native communities through good communication opens possibilities for joint exploration of new areas of enquiry. Currently, there is a trend to integrate traditional scientific practice with other ideological thoughts and practices. For instance, the fusion of religious, social and physical scientific themes is becoming more common in contemporary society. Such interdisciplinary approaches are offering insights into the respective areas of thought. For indigenous communities, this is an appropriate time to inform academic fields based on the integrative knowledge systems found within indigenous communities. Specifically, the holistic thoughts and beliefs of many indigenous communities could serve as a model for the interdisciplinary fusion in the academy. For instance, the traditional knowledge of many indigenous communities is integrative of the physical and metaphysical attributes of life; there is no distinction between spiritual, cultural or physical parameters of the world. Peat (1996) discusses the connections between traditional native knowledge and the implications for modern science.

Inherently, there may be a skeptical view towards such integration. However, it is important to consider how values and norms may fit within a research framework. For instance, the values and norms of a given native community may differ significantly from the values and norms of academics from a given discipline. The values and norms might be applied to the methodology, which might have bearing on the results, thereby preventing replication if the same values and norms are not applied. This theme has been discussed several times at recent gatherings of the Community Forestry and Environmental Research Partnerships (CFERP) fellows. How does the research demarcate social, spiritual and physical relationships to one's project or community? There are no easy answers to this; but the first step is for the researcher to acknowledge what their own perceptions are with respect to such issues. Then, the researcher and community must decide if or how to integrate these attributes within the research project.

In consideration of the historic attrition of traditional knowledge, it may be an objective of native researchers or researchers of native communities to study topics that may assist in the recovery of lost knowledge – for example, research of various attributes of place (climate, vegetation and soils) to restore or better understand place-specific knowledge and language. In some cases, it may be appropriate to look beyond native communities to provide comparative examples of cultural knowledge. For instance, conducting applied research *in situ* may expose the researcher to a greater familiarity of the various parameters involved in developing cultural knowledge of a subject. By considering the pertinence of such parameters to the broader picture of cultural knowledge, the addition of such parameters may complete voids in the community's knowledge base. For instance, fire use among Native Californians is similar to fire use among other indigenous groups globally. Hankins and Petty (2005) have compared Native Californian and Aboriginal Australian community's fire use. Since fire does not occur as a cultural landscape practice within California, the recovery and maintenance of such knowledge has been informed through this comparative research in the Northern Territory, where cultural landscape burning is ongoing.

While the primary objective of such research may emphasize knowledge for the native community, academia also stands to gain from such knowledge. Thus, the road to research in Native topics is two-way. The sometimes difficult task is identifying how to share such knowledge; which format or venue is appropriate; what sorts of information should or should not be shared; how might the information be used; and might that use be detrimental to the community. The repercussions of sharing knowledge should certainly be considered and given with caution. However, we should also remember that many forms of traditional knowledge are communal and are considered the possession of a people, rather than an individual or group of individuals. Certainly, specialists exist within various cultural groups; but the knowledge obtained by the specialists is largely a collective inheritance bestowed by generations of practitioners. At any rate, it is advisable to discuss such matters with community members if any doubt exists as to the dissemination or citing of such knowledge. Similarly, the context for applying some types of knowledge may be relevant to convey.

Conclusion

Boyer (2006, p1) recounts a story about a Laguna Pueblo fifth grade student and his science fair project comparing the benefits of grinding corn with a mill or by hand using a traditional grinding stone. The mill is fast and can grind more corn than a woman can using the stone. But more people contribute by helping when the stone is used, and people get together for songs and stories as the cornmeal is prepared. So if you can make the meal either way, which way makes the stronger community?

Thus, the same question is posed about conventional research and PR. At the end of the day, both will yield information; but which way makes for stronger research and stronger scholars? Which way benefits the community?

There is no question that PR is not for the faint of heart. It is a thought-intensive, people-intensive, labor-intensive and time-intensive methodology. The research itself becomes a social act, and that is important both for control and capacity-building at the local level (where information can have a real effect) and to address past research that has been markedly antisocial in impact, if not intention.

Native communities in the Americas and worldwide have done a remarkable job of protecting knowledge systems and keeping them alive. The decision to share community knowledge must be particularly painful sometimes – almost an act of faith. It is, therefore, important to have research processes that recognize what has been done in the past in the name of science and discovery, and to move towards real partnership and real utility.

For native scholars and communities, there are any number of reasons not to embark on research and not to extend a hand of welcome to someone who says they would like to help. But there are questions to answer, dilemmas to solve and ways to do this that do not entail diminishing a person or a people.

Participatory researchers make room to confront messy issues, ugly history and thorny problems. We think this is especially well suited to creating a transformative future for native researchers and native communities.

References

AISES (American Indian Science and Engineering Society) (2008) *Mission Statement*, www.aises.org

Anderson, M. K. (2005) *Tending the Wild: Native American Knowledge and the Management of California's Natural Resources*, University of California Press, London

Baker, R. (1992) 'The clam "gardens" of Tomales Bay', *News from Native California*, vol 6, no 2, pp28–29

Berkes, F. (2000) 'Rethinking community-based conservation', *Conservation Biology*, vol 18, no 3, pp621–630

BigEagle, D. (1983) 'My grandfather was a quantum physicist', in J. Bruchac (ed) *Songs from This Earth On Turtle's Back*, Greenfield Review, Greenfield Center, New York, NY

Bosk, B. (2004) 'The Ballanco Chronicles', *The New Settler Interview*, no141, pp5–20

Boyer, P. (2006) 'Should expediency always trump tradition?', *Tribal College Journal*, vol 18, no 2, pp12–15

CIBA (California Indian Basketweavers Association) (2007) *Systemic or Social Changes We Are Trying to Achieve*, www.ciba.org/aboutciba.html#change

Gilbert, J. (2001) 'Trodding the circle from Indian community to university research and back', in K. James (ed) *Science and Native American Communities: Legacies of Pain, Visions of Promise*, University of Nebraska Press, Lincoln, NE

Hankins, D. L. (2005) *Pyrogeography: Spatial and Temporal Relationships of Fire, Nature, and Culture*, PhD dissertation, University of California, Davis, CA

Hankins, D. L. and Petty, A. M. (2005) 'A comparison of indigenous pyrogeography in central California, USA, and the Northern Territory, Australia: An ethnoecological primer', in D. L. Hankins (eds) *Pyrogeography: Spatial and Temporal Relationships of Fire, Nature, and Culture*, University of California, Davis, CA

Moller, H., Berkes F., O'Brian Lyver, P. and Kislalioglu, M. (2004) 'Combining science and traditional ecological knowledge: Monitoring populations for co-management', *Ecology and Society*, vol 9, no 3, p2

Peat, F. D. (2002) *Blackfoot Physics: A Journey into the Native American Universe*, Phanes Press, Grand Rapids, MI

Reason, P. and Bradbury, H. (2000) *Handbook of Action Research: Participative Inquiry and Practice*, Sage Publications, London

Schroeder, K. B., Malhi, R. S. and Smith D. G. (2006) 'Opinion: Demystifying Native American genetic opposition to research', *Evolutionary Anthropology*, vol 15, pp88–92

Tuhiwai Smith, L. (1999) *Decolonizing Methodologies*, Zed Books, London, p2

12

Participation, Relationships and Empowerment

Carl Wilmsen, William Elmendorf, Larry Fisher, Jacquelyn Ross, Brinda Sarathy and Gail Wells

WHO DECIDES?

One thing that stands out in the case studies in this book is the quality of the relationships between the community members and professional researchers in each of them. Participation, community, community-based natural resource management (CBNRM), the environment and the economy are all moving targets. Relationships are thus central in shaping the outcomes of engaging with each. While participation is often presented as the key to community capacity-building, it may be that participation has been overemphasized in recent years. Perhaps a more pertinent question to ask is who decides? In every participatory research project, choices need to be made about how the community is defined; who participates in the research, how and when; what should be the focus of observation; what variables should be measured; what methods of data collection and analysis are appropriate; what actions, if any, for social change should be taken; and many other aspects of the project. Ideally, community members and professional researchers treat each other as equal partners in making these decisions together. Doing this successfully rests on establishing trust, reciprocity and credibility – in other words, establishing good relationships.

In the traditional approach to science, the scientist(s) decided. The goal was to avoid forming relationships and to hold the scientist outside of the situation under study, as if he (and the assumption was that scientists were male) were hovering over the playing field below, dispassionately collecting information and recording events as they unfolded. It was assumed that the scientist could lay aside his own biases and become a neutral observer, capable of extracting the essence of things and events and drawing conclusions about them that were universally relevant and applicable.

More than two decades of scholarship in the social sciences have demonstrated that such objectivity is impossible to achieve. All knowledge is shaped by the social backgrounds of the knowledge producers. Culture, class, gender, life experience and other variables all influence the choices made in conducting research and thereby shape substantive knowledge. For the reasons we explained in Chapter 1, participatory research seeks to expand the pool of people who participate in the production of knowledge – in the making of collective social judgments about the situation under study. As the case studies in this book demonstrate, involving community members can improve experimental designs and analyses by bringing in local knowledge, can enhance the involvement of marginalized peoples in natural resource management, and can produce results that are relevant to the lives and livelihoods of the community members collaborating in the research.

The case studies also reveal, however, that realizing these benefits is not automatic. It takes careful work and careful attention to establishing relationships to bring about such promised results. The case studies suggest that areas which commonly present challenges include the quality of participation, ambiguity in defining the community, relationships of power bearing on the research situation, and the capacity of communities to respond to change.

THE QUALITY OF PARTICIPATION

How are relationships and participation interconnected, and what are the implications for the empowerment of communities? The degree and quality of participation are crucial considerations in the practice of participatory research because high levels of participation do not guarantee high-quality involvement. The degree of participation is an issue that must be negotiated among co-researchers in every participatory research project, and the quality of participation must be evaluated on an ongoing basis (Bradbury and Reason, 2003). These measures are necessary for two reasons:

1 Participation is often used as a means of implementing programs and projects designed and controlled by scientists, government officials, non-governmental organizations (NGOs), or other entities external to the affected communities.
2 Sometimes there may be good reasons for community members not to participate.

Many projects are participatory in name only. As Cornwall and Jewkes (1995) have observed, many research projects that claim to be participatory are actually initiated, designed and conducted entirely by scientists. This is a practice that may be considered an abuse of participatory research (Hagey, 1997). Simpson (2000) has similarly protested that researchers in recent years have misled First Nations communities in Canada into believing that their research is participatory when, in fact, it excludes community members from collaborating in many phases of the research, including the application of the research results.

Such charges have serious implications for participatory research. Proponents of participatory research present it as a means of conducting research that will have tangible benefits for the community, and that these benefits accrue through community members' direct participation in the research process. Indeed, we make such claims in many places in this book. Yet, if participation is not meaningful, and if benefits do not accrue to the community, then participatory research will commit the same transgressions of which conventional science is often accused: that the research extracts knowledge from the community to the benefit of the researcher or researchers, and that the community is left unchanged or worse off than it was before.

Researchers and community planners have observed and analyzed the many levels of participation, and associated abuses, since the 1960s. Arnstein's (1969) well-known ladder of participation identified eight levels of participation, from manipulation in which political leaders or others in positions of power 'educate' the public and engineer support for their plans and projects, to citizen control, in which people initiate a project and control its design and management, as well as the application of its results. Pretty (1994, 1995; see Table 12.1) has more recently published typologies nearly identical to Arnstein's. Table 12.1 is a typology derived from his publications.

Typologies such as Arnstein's and Pretty's are useful because they demonstrate that participatory research can be conducted in many different forms for many different motives and goals, and abuses may occur. However, these typologies also suffer from two major disadvantages: they imply that higher levels of participation will lead readily to empowerment and that low levels of participation will inevitably result in extraction or abuse, and they do not account for the common use of conventional field research techniques in participatory research, or for the contributions such techniques can make in the pursuit of empowerment.

While lower levels of participation, lying towards the manipulation end of Arnstein's and Pretty's typologies, fail to generate a sense of ownership, build strong local-level institutions, build leadership skills, reduce conflict and develop other community characteristics considered necessary to sustainable natural resource management, a distinction needs to be made between low levels of participation for maintaining existing relationships of power and low levels of participation driven by community choice or research design. It may be the case that for issues of confidentiality or other reasons, it is best for community members *not* to participate in some phases of a research project. In addition, as Hayward et al (2004) suggest, communities may have very good reasons, such as a lack of time, for declining to participate in research or development projects.

Table 12.1 *A typology of participation in development programs and projects*

Typology	*Characteristics of each type*
1 Manipulative participation	Participation is simply a pretense, with 'people's' representatives on official boards, but unelected and lacking power.
2 Passive participation	People participate by receiving information from agencies about what is going to happen or has already happened. It is a unilateral announcement by agencies without public input. The information being shared belongs only to external professionals.
3 Participation in giving information	People participate by answering questions posed by researchers using questionnaire surveys or similar approaches. People do not have the opportunity to influence proceedings.
4 Participation by consultation	People participate by being consulted or answering questions. External agents define problems and information-gathering processes, and therefore control analysis. This process does not necessarily concede any share in decision-making, and professionals are under no obligation to take on board people's views.
5 Participation for material incentives	People participate by contributing resources (e.g. labor) in return for food, cash or other material incentives. Farmers may provide the fields and labor, but are not involved in experimentation or in the process of learning. It is very common to see this called participation; yet people have no stake in prolonging activities when the incentives end unless the activity makes economic sense or meets other landowner needs. Cost-sharing may improve prolonged activity because of personal investment.
6 Functional participation	People participate by forming groups to meet predetermined objectives related to the project. Such involvement may be interactive and involve shared decision-making, but tends to arise only after major decisions have already been made by external agents. These institutions tend to be dependent upon external initiators and facilitators; but many become self-reliant.
7 Interactive participation	People participate in joint analysis, which leads to action plans and the formation of new local institutions or the strengthening of existing ones. Participation is seen as a right, not just the means to achieve project goals. The process involves interdisciplinary methodologies that seek multiple perspectives and make use of systematic and structured learning processes. These groups take control over local decisions; thus, people have a stake in maintaining initiatives, structures and practices.
8 Self-mobilization	People participate by taking initiatives independently of external institutions in order to change systems. They develop contracts with external institutions for the resources and technical advice that they need, but retain control over how the resources are used.

Note: Used with permission of *World Development*.
Source: Pretty (1994, 1995)

Even when community members do participate, Hayward et al (2004) warn, participation may not always be a social good leading to community empowerment and sustainable outcomes. The degree and quality of participation is affected by any number of variables, and positive social change is by no means assured. Guijt and Shah (1998) point out that the standardization of participatory research techniques and the use of manuals and blueprints for participatory research projects shifts researchers away from context-specific research and encourages them to claim that participation has been successful even in the absence of careful attention to empowerment principles. They point out, for example, that it is often erroneously assumed that participatory rural appraisal (PRA) will automatically take care of gender issues. Sarin (1998) demonstrates that there is nothing automatic about it. She describes a PRA exercise concerning joint forest management in India that did not enable women to speak openly about their issues. Participatory processes and group dynamics may thus provide greater benefits to and reinforce the interests of community members whose social position already accords them greater status and privilege than others in their community (Guijt and Shah, 1998; Cooke and Kothari, 2001).

Yet, moving beyond high-profile community members poses many challenges. As Ballard and Sarathy (see Chapter 8 in this volume) point out, some people may be excluded from participation by the very way in which 'community' is defined. The process of defining concepts such as 'community' may itself be exclusionary because people are uneven in their ability to negotiate how they are represented in the broader discourse or to challenge commonly accepted notions about what constitutes community, skilled labor or even environmental degradation. Moreover, even when a participatory project does include all interested parties, the notion that they will all collaborate convivially on a level playing field is naive. Community members have different levels of legitimacy and different social and economic vulnerabilities. They may have histories with one another that may include personal animosities. All of these bear upon their contributions to participatory processes (Jones, 2003).

The key point for participatory research is that decisions about levels of participation should be made for good reasons, and should be determined in close collaboration with community members. These good reasons will depend upon the context in which participation occurs (Jones, 2003). What distinguishes meaningful from exploitative participation is not whether community members are fully involved in every phase of the research, but whether the research process permits them to negotiate for themselves the level of participation that they desire and that is in the best interest of producing the most trustworthy research results possible. In other words, the answer to the question 'who decides' is empowering for them.

Ambiguity in defining community

Yet, building relationships so that decisions are made jointly between community members and professional researchers is complicated by the diversity of communities and the need to account for varied perspectives in the research. Defining the community is thus unrelentingly political. This has important implications for building relationships between community members and professional researchers.

The problem of who constitutes the community has vexed social scientists and natural resource managers for decades (Machlis and Force, 1988; see also Chapter 4 in this volume). In recent years, scholars studying community-based development and management systems have noted that the concept of community, in its usual sense, hides the diversity within social groupings, obscures power relations and oversimplifies human–environment interactions. Communities are more diverse than use of the concept often implies. Even in small spatially bounded communities that may appear homogeneous to outsiders, differences of gender and social class exist (Agrawal and Gibson, 1999). These differences entail different subject positions, interests and levels of authority among social groups. Kindon (1998), for example, reveals how women in communities in Bali are not equal to men in the work they do or the authority they wield despite prevailing myths among government officials about harmonious relations between the sexes on the island. In addition to concealing diversity, Kindon argues, such myths perpetuate uneven relationships of power.

Relationships between communities and the outside world also contribute to internal diversity. Contemporary patterns of development, settlement and migration between rural communities and urban centers, as well as part-time residence in distant provinces or other countries for work, have created part-time insiders, and also rendered the boundaries of geographic communities difficult to define. Moreover, people who leave to get educated and later return may have difficulty reintegrating within the community, acquiring an ambiguous status of neither insider nor outsider (Guijt and Shah, 1998; see also Chapter 10 in this volume).

This diversity within communities suggests that all homogeneity must be hand picked (Lane and McDonald, 2005). Chapter 5's Tending and Gathering Garden (TGG) is a case in point. By involving community members in the design and implementation of her project, Brawley did what must often be done in participatory research and development: she created a community around her project. Reaching out to many different interests, involving them and bringing in marginalized groups can create a community among social groups that previously may have had little interaction with one another.

On the one hand, creating community in this way is a goal of community development and often of participatory research. As such, it has positive effects such as building human capacity and community agency. On the other hand, the fact that community can be created reveals that community is a process, a moving target along a number of scales involving a number of variables. It is thus the site of political contest and negotiation.

Reflection on the people involved in Chapter 5's project reveals the ambiguity in the use of the term 'community', as well as the politics of the distinction commonly made between communities of place and communities of interest. The people involved in the TGG project include Native Americans who reside in the Cache Creek watershed, or nearby, who are from several different tribal groups. Others involved include local farmers and gravel industry officials. What binds these diverse groups together in this effort is a common interest in managing the Cache Creek watershed in ways from which they all can benefit. In this sense, then, they are a community of interest, albeit one with members who are connected in multiple ways to the local watershed.

Insofar as the individuals involved in the TGG all call the Cache Creek watershed, or some part of it, home, they are also part of a community of place. This suggests that the distinction between communities of place and communities of interest is a false dichotomy. But is it not simply a matter of defining community, in this case, at too large a scale? There are many towns in the Cache Creek watershed; thus, creating community at the watershed scale necessarily entails creating a community of interest because the people involved have attachments to different communities of place.

On the other hand, however, defining 'community' on the watershed scale highlights the fact that people are positioned within intricate webs of spatially connected social interaction, including relationships formed through work, school, play (recreation), worship, political activity, family, and other interests and activities. Each of these may entail interaction with the same or different individuals, or a mixture of them, and each may or may not occur within the same geographically defined space, even at smaller geographic scales. Each individual's 'community' may thus involve relationships with people inside and outside particular geographically defined communities. Indeed, people may consider themselves part of many communities.

By the same token, individuals connect to space in multiple ways, including working the land for profit, cultural reasons or as part of one's hobby; using the land for recreation or spiritual purposes; and simply occupying land as a dwelling place. People get involved in all or some of these activities in different places. They may thus have attachments to multiple places. In the case of public lands, people from different communities come together on the same pieces of land for the same or different purposes. On private lands, what an individual landowner does may adversely affect or benefit his or her neighbor.

Communities of place and interest thus overlap in many ways that belie the dichotomy often drawn between them. People in communities of place have specific interests in local natural resources, and people in communities of interest may have close ties to the places where the natural resources in which they have an interest are located (e.g. partial-year residency; employment involving local natural resources with residence elsewhere; friends, relatives or colleagues who live in those places; love of a particular landscape feature) (London, 2001). Indeed, community is a process that involves interaction among people with attachment to particular places. Or, as Elmendorf and Rios suggest in Chapter 4, it is a place-oriented process with social interaction as its foundation. Place and interest are integral elements of all communities (Wilkinson, 1991). The distinction between communities of place and communities of interest is thus more imagined than real.

Yet, the distinction does have real effects on the abilities of actors occupying different social positions to protect their interests. Each relationship with a place and its natural resources permits people to make certain claims and proscribes them from making others. How people identify themselves, and are identified by others, as belonging to a particular community positions them relative to other groups with different claims on the resource, different sources of legitimacy, and different politically strategic advantages and disadvantages.

The communities of place/communities of interest divide is thus the site of contention and negotiation. People who can claim membership in a community of place use such conceptualizations of community strategically to claim superior rights of participation and access to resources. People in communities of interest may use appeals to the broader good in making such claims. Some environmentalists argue, for example, that privileging place may go against national interests, and therefore interests external to geographically defined communities should be included in policy formulation and management decision-making as well (McCloskey, 1998).

The politics of community play directly into the exclusion of particular groups from engagement in the civic affairs of a community, including natural resource management, as well as from particular community spaces. Ballard and Sarathy in Chapter 8 provide an example of how Latino immigrant workers have been historically excluded from participation in community affairs, as well as from particular public spaces in Oregon's Rogue River Valley. This deliberate social exclusion combined with undocumented legal status, dependence upon kinship and social networks for employment, and a federal agency focus on efficient provision of ecosystem management services effectively bars Latino worker involvement in ecosystem research and management. Ballard and Sarathy argue that Latino forest workers should be involved in ecological studies of natural resources because they can contribute unique insights into the environmental impacts that their use of the resource has.

Yet, involving them in research is challenging. Ballard and Sarathy each followed principles of participatory research in their respective studies, but found that many of the same factors that exclude Latino forest workers from the civic life of communities, as well as from conventional ecological science, also inhibited their involvement in participatory research projects. The workers' undocumented legal status, dependence upon kinship and social networks, and historical experience with prejudice made overcoming distrust and establishing working relationships with them especially challenging. Ballard and Sarathy both found that working through established social service-providing non-profit organizations to be the most efficient way of surmounting these obstacles. Community-based organizations that have gained the trust of local workers and are familiar with and have ties to the academy may thus serve an important function as a liaison between outside researchers and marginalized social groups. Yet working through such organizations may be limiting in that they too have their own agendas that shape the scope of their own social networks. As Ballard observes, despite her close relationship with the Northwest Research and Harvester Association (NRHA), her access to some groups of workers remained limited because her interviewing of labor contractors, crew foremen and government officials inhibited her ability to overcome their distrust of her.

On the other hand, for people who do get involved, participatory research and community development can be powerful exercises in revitalizing communities. As Elmendorf and Rios point out in Chapter 4, African Americans have long been assumed to have little interest in the environment, and this has presented obstacles to their participation in community-based environmental management, as well as in addressing environmental issues more generally. Elmendorf and Rios' case study of community-based planning of parks and gardens in the poverty-stricken Belmont neighborhood of West Philadelphia corroborates other studies in finding that African Americans do, indeed, have deep concerns for environmental matters. They suggest that community-based participatory planning can tap into this reservoir of concerns, knowledge and experience, and nurture a sense of *self in place*, which, they argue, is more important than a sense of place alone. Sandercock (1998) likewise suggests that struggles over urban form are ultimately struggles over belonging. Building community through participatory processes can help to instill or reinforce that sense of belonging in urban residents.

The implication that this has for participatory research is that 'community' is a fluid and contested concept; understandings and representations of local communities are products of negotiation of nation state, local and other interests (Li, 1996). The challenge for participatory research practitioners is to break free of commonly accepted conceptualizations of community and to develop relationships with community members that permit and even encourage joint decision-making even among marginalized groups, and that simultaneously permit critical analysis of all interests in the situation under study.

The latter point introduces a second challenge in the nurturing of relationships with communities, however. A goal of participatory research is to produce knowledge that is relevant to the community. That knowledge must stand up to critical scrutiny or it will be of little use to the community in the long run. Yet, although scientists and community members both value critical scrutiny, their goals, expectations and career incentives often differ when engaging in it. The need to balance rigor and relevance thus introduces tension into the relationship between community members and professional researchers.

BALANCING RIGOR AND RELEVANCE

Balancing rigor and relevance is a central concern of participatory research that is hindered by current divisions in standard research practice. As Argyris and Schön (1991) have observed, if social scientists adhere to prevailing scholarly practices of rigor, they run the risk of producing results that are irrelevant to practitioner demands for usable knowledge, and if they emphasize relevance to practitioners, they risk falling short of current standards of rigor. This apparent dilemma is rooted in the fact that current evaluative criteria in science are founded on the premise of a fundamental separation between basic (enquiries into the fundamental nature of phenomena) and applied (enquiries to support resolution of practical problems) research.

Participatory research is based on the principle that rigor and relevance need not be mutually exclusive. Balancing rigor and relevance requires continuous monitoring of the research process. While many factors are important in this monitoring, encouraging reciprocity, avoiding bias in the research and contributing to theory-building require special attention.

Reciprocity and the quality of knowledge

Wulfhorst et al (see Chapter 2 in this volume) find that the lack of clarity about what constitutes participatory research in the first place defies the development of evaluative criteria. They suggest that three core elements are essential to designing and evaluating participatory research projects. These core elements are degree of community control, reciprocal production of knowledge, and utility and action of outcomes (that is to say, the degree to which the outcomes are useful for taking action in order to effect social change). They echo the three elements identified by many scholars as common to participatory research: participation, knowledge production and social change. Each of these criteria depends upon local context for the way in which it unfolds in any given research project. There are thus no formulaic rules for judging rigor and relevance that apply to every case.

Nevertheless, some means of evaluating the trustworthiness of research findings is necessary. Trustworthiness is the degree to which we can be reasonably sure that research results accurately reflect the situation under study. Participatory research practitioners use a number of techniques for assessing trustworthiness, many of which are also used in conventional research. These techniques include triangulation of methods, checking by peers and participants, inquiry audits (providing enough information about the process that a disinterested third party could evaluate it), assessing the utility of outcomes, and several other measures (Pretty, 1995; Mays and Pope, 2000).

Wulfhorst et al's (Chapter 2) key contribution is to identify reciprocity in the production of knowledge as a core criterion for evaluating participatory research projects. Indeed, several of the chapters in this volume support the conclusion that reciprocity plays a key role in balancing rigor and relevance in the production of knowledge in participatory research. Reciprocity requires mutual respect for other knowledge systems, as well as transparency, trust and credibility. It is central to nurturing relationships between community members and professional researchers.

Mutual respect for other knowledge systems entails acknowledging that there are many different ways of knowing the world of which science is but one. This is not the same as saying that all perspectives are equally valid or all information is equally flawed. Rather, it is simply saying that all knowledge systems are capable of providing valuable insights. Those insights need to be evaluated, however. The assumptions that analysts in any knowledge system, including science, bring to problems, how they define what the problem is, and the way they go about gathering information all influence the knowledge that is produced. Thus, just as scientific findings must be evaluated for their validity and reliability, information produced by non-scientific systems of knowledge must be similarly evaluated (Pretty, 1995).

Mutual respect is crucial to the process of evaluating knowledge and information because Western science is implicated in the history of colonialism, and because its practice can still contribute to social inequities. Western scientific knowledge was an active component of the exercise of colonial power in extracting knowledge and materials from colonized locales for the benefit of Europe (Smith, 1999). Moreover, unequal relationships of power were, and still are, maintained through knowledge production even as oppressed peoples challenge the conditions of their oppression (Bhabha, 1983; Pratt, 1992; Butz and MacDonald, 2001). Feminists, for example, have argued that the traditional practice of science, and the epistemology upon which it is based, employs a double erasure of women: it excludes them from the community of knowers (i.e. scientists) and excludes their life experience as appropriate for study (Sandercock, 1998). Postmodern thinkers, for their part, have emphasized the way in which knowledge/power, in which science is a central element, produces self-disciplining subjects that internalize and reproduce prevailing relationships of power, although at times challenging the exercise of power (Foucault, 1977). While these interpretations are open to debate, postmodernist, postcolonial, feminist and other scholarship have demonstrated the limitations as well as the power effects of the Western model of knowing (Allen, 2001).

Reciprocity in participatory research is crucial to addressing those effects. Reciprocity means making the research process an exchange in which the researchers and their local collaborators share knowledge with one another in open dialogue that includes mutual respect, but also critical evaluation of information, evidence and interpretations. The latter activity is crucial because although participatory research is based on the principle that objective truth is not attainable, clearly some knowledge is more trustworthy than others in understanding the world and in achieving results. Because of the power relations between professional researchers and community members, as well as of the history of ill effects of research on communities, rigorously evaluating information, evidence and interpretations requires open dialogue, agreed-upon rules and trust. Several of this book's chapter authors provide evidence of such requirements.

Ballard et al (see Chapter 9 in this volume) demonstrate this need in reconciling differences in scientific, local and indigenous knowledge. While their chapter clearly shows the utility of local knowledge to ecological research, it simultaneously demonstrates that local knowledge is dependent upon local economic, social and political context. In conducting studies of the effects of harvesting on the floral green salal (*Gaultheria shallon*) in two separate communities on Washington's Olympic Peninsula, they found very different levels of local knowledge of, and experience with, harvesting the shrub. Latino workers in Mason County had intimate knowledge of the resource, which they demonstrated in their descriptions of the response of brush species to management interventions such as tree thinning, and which they applied to establishing Ballard's experimental design. In contrast, the Makah people who participated in the research, while having intimate knowledge of other forest resources, were not as knowledgeable about commercial use of brush species because they had not had recent experience harvesting them for commercial purposes.

Thus, the local knowledge of the Latino forest workers and the indigenous knowledge of Makah tribal members was indelibly shaped by their different relationships with the resource, which were themselves shaped by historical and political economic context. This supports Mosse's (2001) proposition that often what is thought of as local knowledge is really hybrid knowledge produced through the unfolding of uneven relationships of power in research and development processes. Yet scientific knowledge is also shaped by contextual factors. Reconciling differences in the knowledge that people from different knowledge traditions bring to the research process thus requires critically evaluating each contribution. Establishing relationships of reciprocity in this process is crucial to forestalling the misappropriation of local and/or indigenous knowledge, or otherwise contributing to the marginalization of local communities.

Yet, the politics of sharing knowledge is further complicated by local customs and beliefs about the very act of sharing knowledge itself. As Long et al (see Chapter 10 in this volume) observe, in White Mountain Apache communities, sharing knowledge is not always deemed a wise thing. According to the local view, knowledge is powerful and can inflict much damage if misused. Such culturally based prescriptions for restricting access to knowledge until a person, through proper training, is ready for it are counter to the Western scientific disposition favoring free open access to all knowledge. Reciprocity in participatory research thus entails not only mutual respect for different knowledge systems, but also negotiation of what knowledge will be shared, how it will be used, by whom, and to whose benefit (see Chapter 11 in this volume for further detail on the ethics of conducting research in native communities).

The different expectations that research collaborators have for the research also shape the politics of sharing knowledge. Cheng et al (see Chapter 7) succinctly identify the creative tension coursing through the process of monitoring national forest planning because individuals involved and interests represented have different expectations about how they will affect the process and its outcomes (see Table 12.2). They suggest that managing this tension is a fundamental challenge in collaborative processes. They emphasize the need for open and honest communication at all times, clearly defined roles for participants, clearly defined criteria for measuring progress and continuous flexible monitoring of the process. Castellanet and Jordan (2002) corroborate this finding. They report that in a participatory action research (PAR) project in the trans-Amazonian region of Brazil, failure to clearly communicate the different interests and expectations of the different groups collaborating in the research led to feelings of mistrust on all sides and an atmosphere in which information was not always openly shared.

Wilmsen and Krishnaswamy (see Chapter 3 in this volume) argue that the tension between community members and professional researchers is a product of an educational system that fundamentally separates research and community development. Career incentives in the current academic system favor contributions to scholarship over contributions to resolving community development issues. University-based researchers typically want to explore the implications of research results for theory development. While this can produce theory that informs development and conservation practice in different spatial and temporal settings, it may lead to conclusions that are at odds or seem to undercut the goals of community members.

Evaluating the information that each party brings to the participatory research process thus requires acknowledging the different expectations of the parties involved, and adopting procedures that are open, transparent and fair. This means clearly articulating expectations at the outset of research, developing mutually

Table 12.2 *Expectations for research among academics and community members*

Community members	*Academics*
Applicable to local problems	Applicable to other similar situations
Resolving problems	Theory-building
Long-term relationship	Finite end
Respect for tradition	Questioning 'commonsense' notions
Effects on community	Publication

agreed-upon goals and preparing all research collaborators for the eventuality that research conclusions and results may not be to their liking (Firehock, 2003).

Clearly, building trust is important here. If community members do not trust the other parties involved in the research or development project, or suspect for any reason that the process is unfair, they will not accept the findings. Indeed, nearly every treatise on collaboration and community-based natural resource management emphasizes the importance of trust (Smith, 1999; Wondolleck and Yaffee, 2000; Gray et al, 2001; Baker and Kusel, 2003; Krishnaswamy, 2004).

In Chapter 7, Cheng et al add to this already rich literature by showing how trust can be established and destroyed through interactions and relationships across scales, from the local to the national. The United States Forest Service (USFS) is an agency that operates at many scales – from subunits in local ranger districts to entire national forests, from regional offices to the national offices in Washington, DC. Cheng et al (see Chapter 7) report that the community members who participated in the Grand Mesa, Uncompaghre and Gunnison National Forests (GMUG) plan revision process were concerned that the agreements they reached with the local Forest Service officials would not be honored by the national office. Their fears were realized in 2006 when the national office suspended release of the draft forest plan due to concerns that it might counter the Bush administration's oil and gas exploration initiative on public lands. The operation of power across scales thus has implications for establishing and maintaining trust in what may be called 'the scale of trust'.

Wulfhorst et al (see Chapter 2 in this volume) identify another way in which trust bears on the evaluation of information and knowledge. They argue that credibility is a key element in establishing trust, which impinges directly on community capacity-building as well as on the trustworthiness of research findings. Credibility and trust must be mutually established between researchers and community members. Each must trust the other to work together toward agreed-upon goals, and each must feel that the other provides credible information. The question of whose opinion of credibility matters is important here. Wulfhorst et al (Chapter 2) point out that different groups collaborating in a participatory research project may have different views of who is credible and even how their own credibility is perceived by the larger group. They note that while there are no easy prescriptions, it is necessary to address multiple levels of perceptions, including the wider scientific community's perception of the credibility of the research.

Ross et al (see Chapter 5) provide an example of respect and trust that nurtured reciprocity and led to practical benefits to local community members. Their research not only recognized the value of native people's knowledge and gave them credit for it, it also opened opportunities for local native people to have a voice in managing the Cache Creek watershed. In addition, it gave them an educational tool, the Tending and Gathering Garden (TGG), for maintaining and revitalizing their culture on their own terms. It is also important to note that Ross et al's (Chapter 5) project recognized the value of other knowledge systems as well. In developing the TGG, the collaborators also recognized the value of the knowledge of local non-native farmers about restoring soil fertility, as well as the technical knowledge of gravel industry professionals in site restoration. The reciprocity in Ross et al's work thus facilitated outcomes that were clearly relevant to both Native American and non-Native American community members.

A key lesson that the Cache Creek experience holds for participatory research, however, is the importance of historical context in establishing relationships of trust. The efforts of county officials, gravel industry representatives, farmers and environmentalists to end years of conflict over gravel mining and to create a collaborative body for managing the Cache Creek watershed is part of the historical context that predates the creation of the TGG. As Ross et al (Chapter 5) point out, had it not been for this history, the creation of the garden and the associated research would probably not have occurred, certainly not in the form that they took. While circumstances were serendipitous for Brawley and her colleagues, in cases where there is ongoing conflict over natural resource use and management, researchers will have to work from within the local context to build trust and move toward relationships of reciprocity.

Hankins and Ross (see Chapter 11 in this volume) also highlight the importance of reciprocity in relationships between researchers and community members. They argue that reciprocity, conducting research of utility to the community and acknowledging the contributions of native communities to the research at hand, as well as to larger bodies of knowledge more generally, are critical to overcoming the legacy that scientific research has left in native communities. Typically, research conducted in native communities has extracted materials and information without producing direct benefits for the communities. The legacy that this has produced involves failing to recognize and respect native systems of knowledge, typecasting native scholars as 'ethnic' researchers or stereotyping them as environmentally astute or backed by the financial clout of large casinos, and high levels of distrust of the academy in native communities. Hankins and Ross (Chapter 11) suggest that good communication is crucial to nurturing relationships characterized by reciprocity, trust and respect in research, and they describe several elements of strategies for ensuring that good communication occurs.

Establishing relationships of reciprocity requires negotiation. Historical context, the scale of trust, the politics of sharing knowledge, the differing expectations of the research collaborators, the knowledge that each brings to the research process, and other factors determined by local context all bear on building trust, on building community capacity and on the quality of knowledge that the research produces. This highlights the need of researchers to continuously question common assumptions about local, indigenous and scientific knowledge, and to be acutely aware of the politics and power relations of the research situation.

Bias in research

The quality of research results thus depends upon the ongoing negotiations and the decisions that research collaborators make concerning any number of contingencies that arise in response to local context throughout the course of the research. How these decisions are made could introduce bias at any point in the process. This is true of conventional research practices as well; but the concern that participatory research practitioners have for producing research results that are relevant to local community members opens different avenues for the potential introduction of bias.

Consider how the principle of reciprocity bears on relevance. Actively challenging oppression or working for social change of any kind is the fullest extension of the principle of reciprocity. Putting research at the service of the people restructures the traditional relationships of research in which knowledge is extracted from community members, analyzed and applied in different places or in ways in which community members have no interest. Although there are many barriers to overcome, ideally the practice of participatory research empowers community members to apply research results to solving their own problems. Indeed, Wulfhorst et al (see Chapter 2 in this volume) and other scholars have suggested that the utility of the results to the research collaborators be used as a measure of the relevance of the research. Yet, focusing too intently on producing useful results may introduce bias into the research. If the intent of the research is to produce knowledge through a process that improves the situation of the research collaborators, it is legitimate to ask whether the research process itself will yield the results for which community members are looking.

The potential sources of bias in participatory research and conventional research thus differ because of their varying approaches to power relations. The work of all scientists, whether they adopt a participatory or conventional approach, unfolds through power relations, whether they explicitly acknowledge them or not. Conventional research risks introducing bias because its unexamined acceptance of relationships of power may point lines of questioning in directions that interest people in positions of power. Who is granted the legitimacy to determine what questions will be asked, what information is relevant, and what interpretations are valid has much to do with the focus of the investigation. Bias may be introduced simply as the result of the interests of individual researchers, or, at the very worst, through manipulation of funding. On the other hand, participatory research risks introducing bias because its explicit attempt to effect beneficial social change directs research toward questions of interest to community members. The difference is not that one approach is more biased than the other, but rather that the sources of bias are different.[1]

The basic tension between what scientists and community members consider useful, as well as the different expectations that they have for research and knowledge, lies at the core of the varying sources of bias. Scientists want knowledge that contributes to theory. Community members want knowledge that they can apply to solving immediate problems. Participatory research tries to reconcile this difference, and in doing so brings the tension to the fore. Highlighting it is necessary to the continual monitoring that balancing rigor and relevance entails.

Yet, participatory research seeks to move beyond this tension by integrating theoretical and practical knowledge. Characterizing utilitarian and theoretical knowledge as separate establishes a false dichotomy. Utilitarian knowledge, or knowledge in action, is part of the functioning of everyday phenomena and therefore needs to be incorporated within theory in order to build a rich account of the world.

Theory-building

The way in which theory is currently thought about and taught, however, separates it from practical applications. This inhibits the integration of knowledge in action into theory, on the one hand, and mitigates against the direct application of theory to the everyday problems that communities face, on the other. Schön (1995) describes the implications of this separation in the example of a professor who was denied tenure for the familiar reason of failing to produce the requisite number of publications despite having developed an innovative popular module for teaching complex computer algorithms. Schön (1995) observes that what is thought of as theoretically significant prevented this professor from turning his innovative applied research into publishable papers in top journals.

Ideally, theory should be a flashlight[2] – a tool for illuminating underlying practices, processes and interconnections that constitute the root causes of problems. That is to say, it should describe the world well enough to be an adequate explanation in and of itself, *and* so that it can form the basis of addressing the everyday problems that people face. One goal that PR shares with conventional science is to produce better accounts of the world. Unlike conventional science, however, it tries to do that by breaking down traditional barriers between applied and basic research. In practice, this means integrating utilitarian and theoretical knowledge through dialogue and mutual learning. It also means examining and evaluating theory through taking action for social change. In participatory research at its best, action, research and theory are not neatly separable.

Ross et al (Chapter 5) illustrate this point well. Without the action of establishing the Tending and Gathering Garden, there would be no research. Action and research support one another. In contrast to the faculty in Schön's (1995) example, Brawley and her co-researchers turned the project, which was designed to provide practical benefits to local Native American peoples, into a set of researchable questions that will provide results of interest to others engaged in collaborative processes elsewhere.

Cheng et al (see Chapter 7 in this volume) provide a similar example. Their case study is an instance of participatory process-monitoring in which the research is on the process of collaboration itself, and is intended to help improve the collaborative process while it is ongoing. Monitoring, research and collaborative forest planning are interwoven. There is no neat, clean separation of the forest planning process and the study of it. Cheng and his collaborators, too, have combined action and research in a way that can contribute to the building of theory about the operation of collaborative processes.

Joining action and research is hampered by time constraints because it often takes longer than conventional research, however. Researchers, agency officials, community members and others involved in a given research project have different expectations and institutional pressures relating to the length of time spent on research. Researchers may be pressured by the expectations of their institutions to work at a faster pace. Wilmsen and Krishnaswamy (see Chapter 3), for example, observe that the longer time horizons in participatory research are not suited to the time frames of graduate students. The expectation that graduate students should finish their degrees within a few years (two years for MSc students) mitigates against having adequate time to devote to relationship-building. Similarly, players in adaptive management may find it frustrating to have to devote extra time to relationship-building. Cheng et al (Chapter 7), for example, report that Forest Service planning team members became frustrated by the participatory monitoring process in developing the plan for the Grand Mesa, Uncompagre and Gunnison National Forests because of the time it took away from what they considered their primary responsibility of writing assessments and the plan itself. Federal officials were not alone in feeling frustrated about time. Community members involved in the forest planning process also grew weary of the time spent on relationship-building and were impatient with the un-timeliness of delivered reports and other research products.

Yet, it is time consuming to conduct research that is relevant to community members and that includes a broad spectrum of interests and perspectives. As Wilmsen and Krishnaswamy (Chapter 3) point out, in limited-resource communities it may be necessary to devote a great deal of time and effort to community organizing before reaching a point at which the community is ready to even develop a research question. This may involve time commitments and activities at odds with the expectations of funders and others involved, which may, in turn, lead to conflict between people who see value in spending more time in relationship-building, and people who are anxious to see completed products such as reports. While community organizing may be consistent with participatory research's goal of social change, tools and strategies need to be developed for organizing communities in ways that encourage their meaningful involvement in designing and conducting research that is both relevant and rigorous in a timely fashion.

Integrating utilitarian and theoretical knowledge requires managing the creative tension that Cheng and his colleagues (Chapter 7) identify as being central to collaborative processes. This tension stems from different expectations, assumptions and perspectives that the co-researchers have about roles, responsibilities and desired outcomes. Managing this tension requires negotiation, relationship-building and continual monitoring for sources of bias. It is a process that is imbued with power.

POWER RELATIONS

Nurturing relationships of trust and reciprocity between community members and professional researchers, building community capacity and incorporating local and indigenous knowledge within research are all processes that unfold through variously negotiating, confronting and supporting relationships of power. In promoting social change, participatory research practitioners engage power relations on many levels: those between the professional researcher and community members, those within the community and those between the community and external entities. Proponents of participatory research claim that participatory research's emphasis on effecting democratic research processes will build capacity within groups and communities for more democratic decision-making and management practices. While this may be true in some circumstances, it is by no means assured. Prevailing relationships of power apply strong pressure on people to maintain the status quo. This is evident in issues that arise in incorporating indigenous or local knowledge within research, in the need to continually confront power relations, in the rigidity of structures of legal authority, and in barriers to including representatives of all interests in the research.

Negotiating the inclusion of local or indigenous knowledge in research is entangled in relationships of power. Long et al (see Chapter 10 in this volume) examine the power dynamics between Western scientific knowledge and the indigenous knowledge of the White Mountain Apaches. They observe that members and employees of the White Mountain Apache tribe have incorporated Western scientific knowledge within their problem-solving through several different pathways. It is not outside knowledge *per se* that creates problems. Rather, problems arise with how it is brought into the community, how it is applied and who the primary beneficiaries are. The core of conflicts between different knowledge systems is the question of who has the power to control the application of knowledge as well as to distribute the benefits. Through various policies, tribal leaders have tried to control the terms under which outside knowledge is brought into their communities.

The impacts of Western science on native and other communities is well documented in the academic and popular literature (Deloria, 1969; Starn, 1986; Trinh, 1989; Guha, 1997; Smith, 1999), as well as in song (i.e. Westerman's 'Here Come the Anthros'). Long et al (Chapter 10) give a good example of the effect the encounter between Western science and traditional knowledge can have on social relations within the community. When the tribal government hires or advances tribal members who have earned college degrees, without having an equivalent process for members who have acquired traditional knowledge, it can reinforce the impression that outside education is a means of self-advancement and that traditional knowledge is not as highly valued.

Long et al (Chapter 10) suggest that PR forces examination of power relationships more openly and directly. Involving community members entails a dialogue in which all knowledge is respected and evaluated as to its relevance and value to the situation under study. This brings the critical tensions between insider and outsider knowledge to the fore and helps to cultivate shared understandings across cultures. It moves beyond traditional approaches to community development, including technology transfers and conventional education of community members, by offering a means of translating between worldviews.

Yet, facilitating relationships in which this translation can occur is not easy. For empowerment to occur, participants in research and development must actively challenge oppression. Engaging community members in an egalitarian research process does not automatically translate into more egalitarian social relations (Crawley, 1998) or reduced conflict. Even when the intent is to make power relations within a community more egalitarian, communities may fall back on existing relationships of power when applying the results, or taking action in the course of the research (Martin and Lemon, 2001; Schafft and Greenwood, 2002). Moreover, the push for participation itself may prevent the use of other legitimate decision-making processes or methods better suited to the particular situation (Cooke and Kothari, 2001).

Consider, for example, Cheng et al's (Chapter 7) participation in the forest planning process on the Grand Mesa, Uncompaghre and Gunnison National Forests. In this case – indeed, in any case concerning national forests – the Forest Service cannot relinquish its decision-making authority under current law (Wondolleck and Yaffee, 2000). The participatory planning process, which was initiated by the Forest Service, thus maintains the basic structure of authority. This means that meaningfully involving community members in forest planning or other management practices and processes depends upon the good will, vision, skill and career incentives of agency officials. Participatory research is thus vulnerable to the shifting fashions, directives and political will within entities, such as government agencies, with legally established decision-making authority.

Cumming et al (see Chapter 6 in this volume) provide an example of how people in governing bodies seek to protect their interests even in the face of research results produced through a highly participatory, democratic research process. Cumming et al engaged residents of predominantly rural Macon County, North Carolina, in a participatory process to assess their experience with, and attitudes toward, the significant environmental and social change they were facing due to heavy in-migration of people from urban centers. The in-migration has led to significant construction of private residences in the wooded hills and on the ridge tops of the county, raising concerns about environmental damage, aesthetic changes in the landscape and threats to the local way of life. Although the research showed strong support for more stringent land-use controls, the County Planning Board, many of the members of which are aligned with real estate developers and property rights interests, refused to enact any land-use controls. They argued that the research findings were not representative of the views of the majority of Macon County residents. Cumming et al (Chapter 6) have implemented a sample survey of county residents to test how representative the findings of their participatory research are of the views of the majority of county residents. They observe that a less participatory study – the survey – was needed to translate their findings into a research framework that county commissioners would recognize as legitimate.

Community members (in Macon County and elsewhere) are not completely powerless. They can appeal to higher authorities, invoke laws or customs that stipulate their fair treatment or receipt of benefits, bring lawsuits, claim the moral higher ground, engage in civil disobedience, and engage in other acts of resistance or advocacy. Their participation in natural resource management therefore involves, at different times and over different issues, negotiation, contestation and compromise. In this process the discourses and practices of domination and resistance mutually shape one another (Sharp et al, 2000). Through co-opting oppositional claims, hiring activists and other means, dominant interests change their own discourses and practices to accommodate the views of the opposition, albeit in a watered-down or some other 'safer' form. By the same token, to make their advocacy resonate with the broader public, activists must appeal to predominant ideas, myths and values, while simultaneously challenging them (D'Anjou and Van Male, 1998) or using parts of them strategically (Pulido, 1998; Butz and MacDonald, 2001; Wilmsen, 2007). In accomplishing this contradictory task, they incorporate discourses and practices of domination within the discourses and practices of resistance.

This has two implications for research. First, research processes, including the incorporation of local and/or indigenous knowledge, research design and data collection, are negotiated within the confines of prevailing relationships of power. While this is true of both conventional research and participatory research, participatory research practitioners approach this negotiation self-consciously and embrace it as part of the practice. Participatory research produces collective social judgments about the situation under study, and collectively arriving at these judgments necessitates confronting, negotiating or submitting to power. Power thus shapes the questions that are asked, the information that is collected and analysis of the results even when the research has emancipation of marginalized people as a goal. Second, it raises questions about the role of the professional researcher in advocacy. At what point does putting research at the service of the people and building their capacity to challenge prevailing relationships of power become advocacy? Research collaborators must continually guard against allowing their advocacy to compromise the research.

Power relations within the community can affect participation and, hence, the knowledge shared. Recall Ballard et al's (Chapter 9) work on the Makah reservation discussed earlier, in which tribal members were asked to participate in research on a specific use of a shrub species with which they had little experience. Because the Makah project was initiated by resource management professionals in the tribe as an experiment, it was a case of more powerful people within the tribe setting in motion a project that, while intended to benefit all tribal members, was not 'owned' by them. Not having as much to contribute or to gain from the project, tribal members did not participate to the extent or respond as enthusiastically to the research as had the forest workers in Mason County. This suggests, as Ballard et al (Chapter 9) conclude, that the characteristics of a community that would, on the surface, seem to be conducive to conducting participatory research – a place-based community with long occupancy of the land, formal avenues for community members to participate in natural resource management, and internal and external support for the research and development of natural resources – do not necessarily lead to successful participation in every case.

Even when research is designed in collaboration with a variety of different interests within a community, institutional structures, barriers of race, class and gender and limitations in the social networks of the participants in the research may circumscribe who is involved in the research. In addition, even when full, meaningful involvement of community is attained, there is no guarantee that sustainable management of natural resources or more egalitarian social arrangements will be achieved. Recent research in the social sciences as well as in ecology suggests that while local communities may be good land managers under certain conditions, there is no simple direct relationship between common assumptions about communities and the quality of their stewardship of natural resources (Agrawal and Gibson, 1999; Castellanet and Jordan, 2002). The assumption of a homogeneous stable community managing its resources sustainably over the long term – an assumption upon which community-based natural resource management is founded – is problematic because neither community nor the environment are static or stable (Leach et al, 1999).

CONCLUSION

As the chapters in this volume demonstrate, participatory research is an approach for conducting research in more egalitarian ways in the midst of uncertainties in environmental and community dynamics. Participatory research can produce better knowledge, build capacity among community members and promote positive social change. It can also produce more nuanced understandings of processes of environmental change (Castellanet and Jordan, 2002). In order to achieve these results, however, researchers must assess each situation thoroughly and guard against relying on simple assumptions about what constitutes community and what constitutes participation. The ambiguities in the concept of community indicate that romanticizing community risks empowering communities to manage the environment when they do not have the capacity for doing so. As Cleaver

(2001) points out, in advocating participation we risk going from one extreme of assuming that we (the experts) know best to they (the community) knows best. On the other hand, the difficulties in ensuring meaningful egalitarian participation indicate that research practice, even in the guise of participatory research, can disempower communities and maintain or exacerbate social inequities.

The point of participatory research is to avoid both extremes and, instead, nurture a relationship in which the terms of the research and the ways in which the results will be applied are jointly decided among the professional researchers and the people directly connected to the situation under study. Questions of who participates and how, who is the community, what is the researcher's role in the community, what different expectations the co-researchers have, and who will benefit and how from the research need to be worked out anew with every research project. Through establishing trust, building relationships of reciprocity and maintaining a self-conscious awareness of the effects of research on the research community, a relationship that facilitates rigorous, relevant community-based research may be created.

Notes

1 Practitioners of both approaches have developed techniques for mitigating the effects of bias on their research results (see Lincoln and Guba, 1985; Kirk and Miller, 1986; Pretty, 1995; Bradbury and Reason, 2003).
2 We are grateful to Louise Fortmann for this metaphorical expression of the role of theory in social change.

References

Agrawal, A. and Gibson, C. (1999) 'Enchantment and disenchantment: The role of community in natural resource conservation', *World Development*, vol 27, no 4, pp629–649

Allen, T. F. H., Tainter, J. A., Pires, J. C. and Hoekstra, T. W. (2001) 'Dragnet ecology – "Just the facts, ma'am": The privilege of science in a postmodern world', *BioScience*, vol 51, no 6, pp475–485

Argyris, C. and Schön, D. A. (1991) 'Participatory action research and action science compared', in W. F. Whyte (ed) *Participatory Action Research*, Sage, Newbury Park, CA

Arnstein, S. R. (1969) 'A ladder of citizen participation', *American Institute of Planners Journal*, vol 35, no 4, pp216–224

Baker, M. and Kusel, J. (2003) *Community Forestry in the United States: Learning from the Past, Crafting the Future*, Island Press, Washington, DC

Bhabha, H. K. (1983) 'Difference, discrimination, and the discourse of colonialism', in F. Barker, P. Hulme, M. Iversen and D. Loxley (eds) *The Politics of Theory*, University of Essex, Colchester, UK

Bradbury, H. and Reason, P. (2003) 'Issues and choice points for improving the quality of action research', in M. Minkler and N. Wallerstein (eds) *Community-based Participatory Research for Health*, Jossey-Bass, San Francisco, CA

Butz, D. and MacDonald, K. I. (2001) 'Serving sahibs with pony and pen: The discursive uses of "Native authenticity"', *Environment and Planning D: Society and Space*, vol 19, no 2, pp179–201

Castellanet, C. and Jordan, C. F. (2002) *Participatory Action Research in Natural Resource Management: A Critique of the Method Based on Five Years' Experience in the Transamazônica Region of Brazil*, Taylor and Francis, New York, NY

Cleaver, F. (2001) 'Institutions, agency and the limitations of participatory approaches to development', in B. Cooke and U. Kothari (eds) *Participation: The New Tyranny?*, Zed Books, London

Cooke, B. and Kothari, U. (eds) (2001) *Participation: The New Tyranny?*, Zed Books, London

Cornwall, A. and Jewkes, R. (1995) 'What is participatory research?', *Social Science and Medicine*, vol 41, no 12, pp1667–1676

Crawley, H. (1998) 'Living up to the empowerment claim?: The potential of PRA', in I. Guijt and M. K. Shah (eds) *The Myth of Community: Gender Issues in Participatory Development*, Vistaar Publications, New Delhi

D'Anjou, L. and Van Male, J. (1998) 'Between old and new: Social movements and cultural change', *Mobilization: An International Journal*, vol 3, no 2, pp207–226

Deloria, V. J. (1969) *Custer Died for Your Sins*, Avon, New York, NY

Firehock, K. (2003) *Protocol and Guidelines for Ethical and Effective Research of Community Based Collaborative Processes*, www.cbcrc.org/documents.html, accessed 7 September 2007

Foucault, M. (1977) *Discipline and Punish: The Birth of the Prison*, Vintage Books, New York, NY

Gray, G. J., Enzer, M. J. and Kusel, J. (2001) 'Understanding community-based forest ecosystem management: An editorial synthesis', *Journal of Sustainable Forestry*, vol 12, no 3/4, pp1–23

Guha, R. (1997) 'The authoritarian biologist and the arrogance of anti-humanism: Wildlife conservation in the third world', *The Ecologist*, vol 27, no 1, pp14–20

Guijt, I. and Shah, M. K. (1998) 'Waking up to power, conflict, and process', in I. Guijt and M. K. Shah (eds) *The Myth of Community: Gender Issues in Participatory Development*, Vistaar Publications, New Delhi

Hagey, R. S. (1997) 'The use and abuse of participatory action research', *Chronic Disease in Canada*, vol 18, no 1, pp1–4

Hayward, C., Simpson, L. and Wood, L. (2004) 'Still left out in the cold: Problematising participatory research and development', *Sociologia Ruralis*, vol 44, no 1, pp95–108

Jones, P. S. (2003) 'Urban regeneration's poisoned chalice: Is there an *impasse* in (community) participation-based policy?', *Urban Studies*, vol 40, no 3, pp581–601

Kindon, S. (1998) 'Of mothers and men: Questioning gender and community myths in Bali', in I. Guijt and M. K. Shah (eds) *The Myth of Community: Gender Issues in Participatory Development*, Vistaar Publications, New Delhi

Kirk, J. and Miller M. L. (1986) *Reliability and Validity in Qualitative Research*, Sage, Newbury Park, CA

Krishnaswamy, A. (2004) 'Participatory research: Strategies and tools', *Practitioner: Newsletter of the National Network of Forest Practitioners*, no 22, pp17–22

Lane, M. B. and McDonald, G. (2005) 'Community-based environmental planning: Operational dilemmas, planning principles and possible remedies', *Journal of Environmental Planning and Management*, vol 48, no 5, pp709–731

Leach, M., Mearns, R. and Scoones, I. (1999) 'Environmental entitlements: Dynamics and institutions in community-based natural resource management', *World Development*, vol 27, no 2, pp225–247

Li, T. M. (1996) 'Images of community: Discourse and strategy in property relations', *Development and Change*, vol 27, pp501–527

Lincoln, Y. S. and Guba, E. G. (1985) *Naturalistic Inquiry*, Sage, Beverly Hills, CA

London, J. K. (2001) *Placing Conflict and Collaboration in Community Forestry*, PhD thesis, University of California, Berkeley, CA

Machlis, G. E. and Force, J. E. (1988) 'Community stability and timber-dependent communities: Future research', *Rural Sociology*, vol 53, no 2, pp220–234

Martin, A. and Lemon, M. (2001) 'Challenges for participatory institutions: The case of village forest committees in Karnataka, South India', *Society and Natural Resources*, vol 14, no 7, pp585–597

Mays, N. and Pope, C. (2000) 'Qualitative research in health care: Assessing quality in qualitative research', *BMJ*, vol 320, no 1, January, pp50–52

McCloskey, M. (1998) 'Local communities and the management of public forests', *Ecology Law Quarterly*, vol 25, no 4, pp624–629

Mosse, D. (2001) '"People's knowledge", participation and patronage: Operations and representations in rural development', in B. Cooke and U. Kothari (eds) *Participation: The New Tyranny?*, Zed Books, London

Pratt, M. L. (1992) *Imperial Eyes: Travel Writing and Transculturation*, Routledge, London

Pretty, J. N. (1994) "Alternative systems of inquiry for sustainable agriculture", *Institute of Development Studies (IDS) Bulletin*, vol 25, no 2, pp37–48

Pretty, J. N. (1995) 'Participatory learning for sustainable agriculture', *World Development*, vol 23, no 8, pp1247–1263

Pulido, L. (1998) 'Ecological legitimacy and cultural essentialism: Hispano grazing in northern New Mexico', in D. Peña (ed) *Chicano Culture, Ecology, Politics: Subversive Kin*, University of Arizona Press, Tucson, AZ

Sandercock, L. (1998) *Toward Cosmopolis: Planning for Multicultural Cities*, John Wiley and Sons, Chichester, UK

Sarin, M. (1998) 'Community forest management: Who's empowerment...?', in I. Guijt and M. K. Shah (eds) *The Myth of Community: Gender Issues in Participatory Development*, Vistaar Publications, New Delhi

Schafft, K. A. and Greenwood, D. J. (2002) 'The promise and dilemmas of participation: Action research, search conference methodology and community development', Paper submitted to the 65th Annual Meeting of the Rural Sociological Society, Chicago, IL

Schön, D. A. (1995) 'Knowing-in-action: The new scholarship requires a new epistemology', *Change*, vol November/December, pp27–34

Sharp, J. P., Routledge P., Philo C. and Paddison R. (eds) (2000) *Entanglements of Power: Domination/Resistance*, Routledge, London

Simpson, L. (2000) 'Aboriginal peoples and knowledge: Decolonizing our processes', *The Canadian Journal of Native Studies*, vol XXI, no 2, pp137–148

Smith, L. T. (1999) *Decolonizing Methodologies: Research and Indigenous Peoples*, Zed Books, New York, NY

Starn, O. (1986) 'Engineering internment: Anthropologists and the War Relocation Authority', *American Ethnologist*, vol 13, pp700–720

Trinh, T. M.-H. (1989) *Woman, Native, Other: Writing Postcoloniality and Feminism*, Indiana University Press, Bloomington, IN

Wilkinson, K. P. (1991) *The Community in Rural America*, Greenwood Press, Westport, CT

Wilmsen, C. (2007) 'Maintaining the environmental/racial order in northern New Mexico', *Environment and Planning D: Society and Space*, vol 25, no 2, pp236–257

Wondolleck, J. M. and Yaffee, S. L. (2000) *Making Collaboration Work: Lessons from Innovation in Natural Resource Management*, Island Press, Washington, DC

Index

Page numbers in *italics* refer to figures, tables and boxes